The Islands of Benoît Mandelbrot

Fractals, Chaos, and the Materiality of Thinking

Nina Samuel,
Editor

Published by the Bard Graduate Center: Decorative Arts, Design History, Material Culture
Distributed by Yale University Press, New Haven and London

This catalogue is published in conjunction with the exhibition *The Islands of Benoît Mandelbrot: Fractals, Chaos, and the Materiality of Thinking* held at the Bard Graduate Center: Decorative Arts, Design History, Material Culture from September 21, 2012 through January 27, 2013.

Curator of the Exhibition and Editor:
Nina Samuel
Head of Focus Gallery:
Ivan Gaskell
Project Coordinator:
Ann Marguerite Tartsinis
Coordinator of Catalogue Photography: Alexis Mucha
Copy Editor: Adam Lehner
Catalogue Design and Production:
Laura Grey with Ben Tuttle
Media Team:
Han Vu and Kimon Keramidas
Exhibition Designer:
Ian Sullivan

Chief Curator and Executive Editor of Gallery Publications:
Nina Stritzler-Levine

Published by the Bard Graduate Center: Decorative Arts, Design History, Material Culture
New York, NY

Unless otherwise noted: Photographs Han Vu, Bard Graduate Center.

Exclusive trade distribution by Yale University Press, New Haven and London
ISBN (Yale University Press): 978-0-300-18643-7

Library of Congress Cataloging-in-Publication Data

The islands of Benoît Mandelbrot : fractals, chaos, and the materiality of thinking / Nina Samuel, editor ; Matthias Bruhn, Jan von Brevern, Juliet Koss, Margarete Pratschke, Wladimir Velminski.
pages cm

This catalogue is published in conjunction with the exhibition The Islands of Benoît Mandelbrot: Fractals, Chaos, and the Materiality of Thinking held at the Bard Graduate Center: Decorative Arts, Design History, Material Culture from September 21, 2012 through January 27, 2013.
Includes bibliographical references and index.
ISBN 978-0-300-18643-7 (pbk.)
1. Visual communication in science–Exhibitions. 2. Mandelbrot, Benoit B.–Exhibitions. I. Samuel, Nina, curator, editor of compilation. II. Bard Graduate Center: Decorative Arts, Design History, Material Culture.
Q223.I85 2012
514'.742–dc23

2012017043

Printed by GHP, West Haven, Connecticut

Cover: Benoît Mandelbrot and Alan Norton, programmer. Quarternion, 1983. Computer-generated prints on photographic paper. Collection Aliette Mandelbrot.

Frontispiece: Benoît Mandelbrot and Mark R. Laff, programmer. Investigations in the complex plane of iterated polynomials, series of over 150, 1979. IBM. Computer-generated print on photographic paper. Collection Aliette Mandelbrot.

Contents

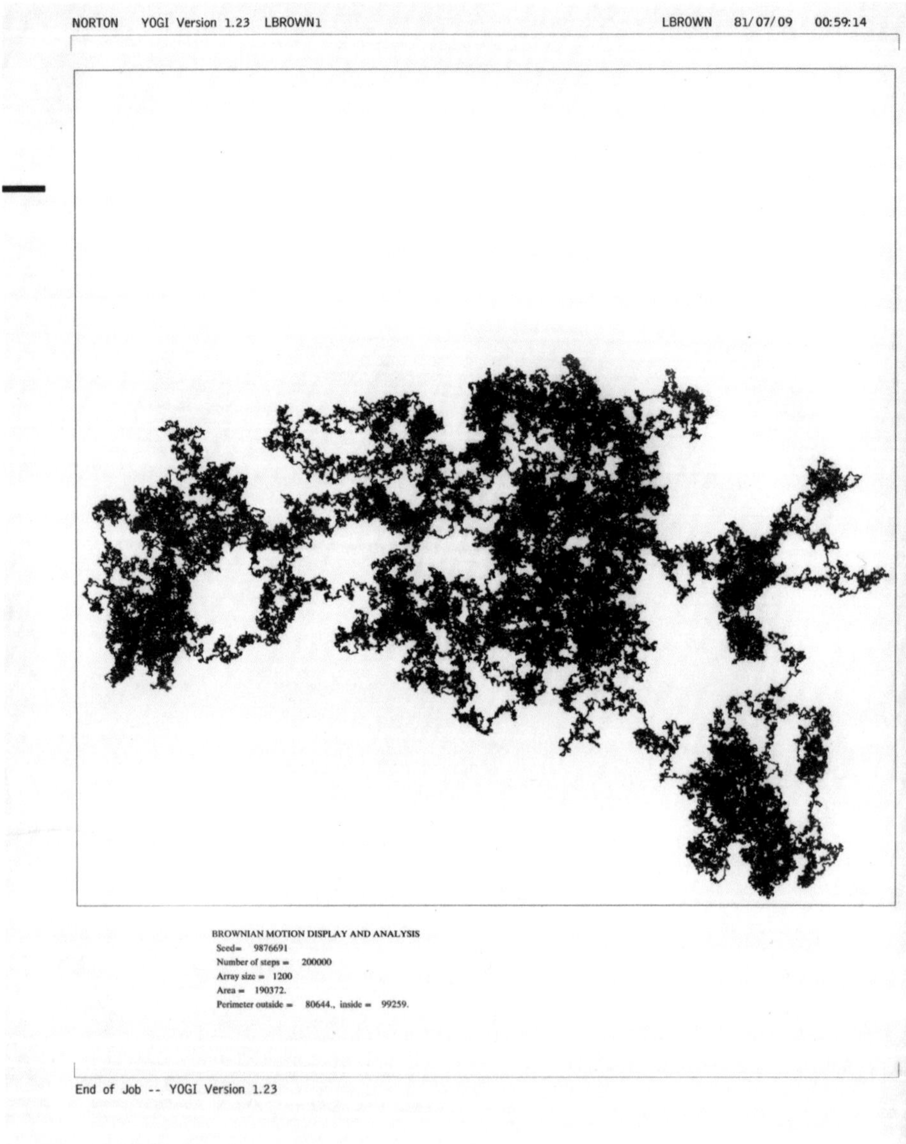

Benoît Mandelbrot. Brownian motion, 1981.
Computer-generated prints. Collection Aliette
Mandelbrot.

Foreword

This groundbreaking volume explores the fascinating hand-drawn and computer-generated images created by the notable mathematician Benoît Mandelbrot (1924–2010) in his pioneering studies in the realm of fractal geometry. Perhaps best known for his work related to chaos theory, Dr. Mandelbrot produced extraordinary works on paper whose resonatingly luminous palettes and distinctive graphic compositions may already be familiar to readers and exhibition visitors, having developed a cult-like following that derives from diverse domains, including the film *Star Trek*. Here, for the first time, the Bard Graduate Center has the great privilege of publishing a special selection of Dr. Mandelbrot's renderings, one that should open up a new path of inquiry in this intriguing field. Exclusively black and white, these beautiful and mysterious images appear to have come from the movement of the hand, seemingly free form, in a way that one might associate with the worlds of automatism and Surrealism rather than that of the computer. Together with the interpretation and contextualization of Dr. Mandelbrot's work and the smaller selection of renderings and drawings by other scientists brought together here, these images suggest unexpected links between the poles of art and science.

Organized in conjunction with our Focus Gallery exhibition, *The Islands of Benoît Mandelbrot: Fractals, Chaos, and the Materiality of Thinking* is also remarkable for having been curated by Nina Samuel, a visiting member of our faculty from Das Technische Bild, a research group at the Hermann von Helmholtz-Zentrum für Kulturtechnik in Berlin that was established by Professor Dr. Horst Bredekamp in 2000. We embarked on this important initiative at the instigation of Dean Peter Miller, and it is indicative of our commitment to fostering collaborative partnerships with other research institutes. Both our curriculum and exhibition practices have been enriched by our connection with Das Technische Bild and its contribution to the formation of this project. Largely based on Professor Samuel's doctoral dissertation, the curatorial thinking on display here came about in a series of seminars that were held at the Center during the past academic year.

Matthias Bruhn and Margarete Pratschke (also of Das Technische Bild) and scholars Wladimir Velminski, Jan von Brevern, and Juliet Koss have joined Professor Samuel in contributing essays that are perfectly in sync with the interdisciplinary nature of the material. I want to thank them for their research efforts and thoughtful insights.

Overall, the realization of this publication and the exhibition that gave rise to it would not have been possible without the assistance of Aliette Mandelbrot, Dr. Mandelbrot's widow. She greatly facilitated Professor Samuel's research efforts and teaching by generously allowing her access to an extensive number of images and agreeing to loan many of these to the Bard Graduate Center. Moreover, Mrs. Mandelbrot authorized access to Dr. Mandelbrot's private study in Cambridge, Massachusetts, an opportunity that had direct implications for the interpretation and selection of works in the exhibition.

This project also reveals the primacy of images in the work of Edward N. Lorenz, a mathematician whose fame derives from his contributions to chaos theory. The Library of Congress provided important opportunities for research and agreed to be a lender of the Lorenz materials in their collection. In addition, we secured loans from Florian Grond, Thomas Kienzl, Gabriele Engelhardt, and Otto E. Rössler in Germany. Mark R. Laff, who worked as a programmer with Dr. Mandelbrot, agreed to not only be a lender but also to kindly participate in Professor Samuel's seminar.

Russell Stockman accomplished the challenging job of translating the German-language manuscripts with great expertise and sensitivity. I would like to thank Adam Lehner for the superb copyediting of this publication. The Center's very talented art director, Laura Grey, created the wonderful graphic design.

This project benefitted substantially from the work of the Gallery staff. In her role as Acting Head of the Focus Gallery, Nina Stritzler-Levine worked with Professor Samuel in the early stages of the project, accompanying her to Cambridge to secure the loans from Mrs. Mandelbrot. When Ivan Gaskell became Head of the Focus Gallery in January, he directed the Focus Gallery team. Ann Marguerite Tartsinis, assistant curator, served as the project coordinator, offering invaluable and ongoing assistance. Alexis Mucha, coordinator of catalogue photography, adeptly procured the illustrations and collaborated with Nina Samuel to formulate the captions. Olga Valle Tetkowski, manager of curatorial projects, secured both institutional and private loans, while registrar Eric Edler coordinated the delivery of the Mandelbrot archive material from Cambridge. Exhibition designer Ian Sullivan worked closely with Professor Samuel to create a display that could evoke the unique aesthetic and meaning of the exhibition's content. Han Vu, digital technology designer, created the films for the project and Kimon Keramidas participated as part of the Focus Gallery team. I am proud to acknowledge BGC students Andrew Goodhouse, Zahava Friedman-Stadler, and Elizabeth St. George: They work in the exhibitions office and assisted with many tasks associated with the production of this catalogue. Lorraine Bacalles together with Joseph Ahearn provided administrative assistance that enabled Professor Samuel to come to the Center in New York. During Dean Miller's sabbatical, Dean Elena Pinto Simon and chair of academic programs Amy Ogata helped coordinate this international endeavor.

I want to thank all of the faculty, administration, and staff of the BGC for upholding the highest quality standards that distinguish this project and all of the Center's activities.

—Susan Weber
Director and Founder
Iris Horowitz Professor in the History of Decorative Arts

Introduction Nina Samuel

I was struck . . . by the way in which many aspects of laboratory practice could be ordered by looking not at the scientist's brain (I was forbidden access!), at the cognitive structures (nothing special), nor at the paradigms (the same for thirty years), but at the transformation of rats and chemicals into paper.[1] *—Bruno Latour*

Chaos and the Order of Things: Material Remains of a Workplace

Cambridge, Massachusetts, fall 2011, the abandoned workplaces of the mathematician Benoît Mandelbrot: a room off one of the neutral gray corridors in the IBM Watson Research Center and a private workroom in a nearby building (fig. 1), both located near Harvard and MIT. A few boxes and filing cabinets are closed, others can no longer be shut owing to the abundance of paper. There is no inventory, and access time is limited. This is no archive. Here it is chance that determines what one sees—or, as Mandelbrot put it in 2008: "The amount of stuff is phenomenal. And it is completely random what was put in boxes and what was put in shelves."[2] The landscape of documents is so vast and impenetrable that it would be impossible to comprehend it at a glance even if there were order—and this is precisely what tempts one to explore it. Its elements, the "things" in this private workplace, will soon be transported to Stanford University, catalogued, and made available to the public. But for now, a life's work is here lying in boxes and files.

The idiosyncratic arrangement found in Mandelbrot's offices developed organically through years of work. It is a whole with a particular character that institutional archiving will destroy, the effect of the mathematician's restless activity in moving objects and papers here and there, an essential element of his daily course of reflection, research, and experimentation. The result

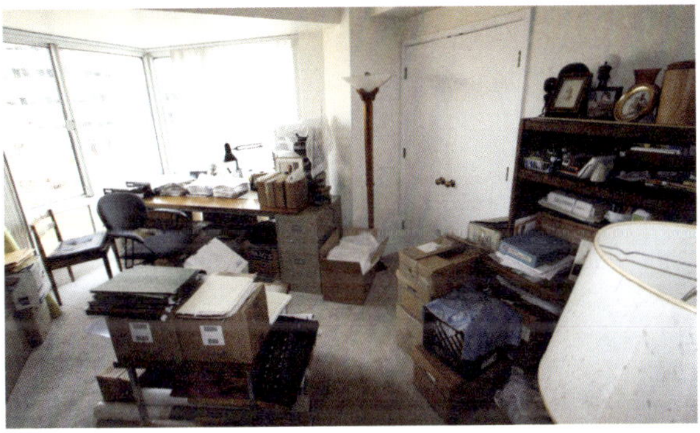

Fig 1. Benoît Mandelbrot's office, Cambridge, Massachusetts, 2011.

is an indiscriminate mixture of personal history and important moments in the history of technology, ideas, and images.

Exploring the private objects, photographs, and working materials in this office is like wandering through the mathematician's brain, witnessing the ephemeral traces of his thought processes. Pointing to the desk at the window, Mandelbrot's widow, Aliette, explains: "Here is where fractals were invented." This is the desk that had once stood in his office at IBM in Yorktown Heights, New York. Mandelbrot worked and did his research there from 1958 until the couple moved to Cambridge in 2007. Portraits and press photographs are distributed throughout the space, mixed with mementos given him by admirers, including a large embroidered picture from Japan (fig. 2). Mandelbrot's public acclaim was enormous, wholly unprecedented in the field of mathematics. But the fact that his imagery became so famous outside the mathematical community was a curse as well as a blessing; as the mathematician Gregory F. Lawler stated, "It led to an overestimation of his contributions outside mathematics and to an underestimation inside the community."[3] Mandelbrot created powerful pictures that became symbols of an era—even a turning point in time—as well as icons of a theory.

The layers of time gathered in these boxes and drawers reflect a scholar's life, one distinguished by interdisciplinary diversity. After studying mathematics in France and two brief stints doing post graduate work in the United States (in aeronautics at Caltech and the Institute for Advanced Studies at Princeton), Mandelbrot emigrated from Europe, accepting a post at IBM. He abandoned a purely academic career because, he maintained, "IBM was far better than any university department in either country."[4] During the 1970s, he

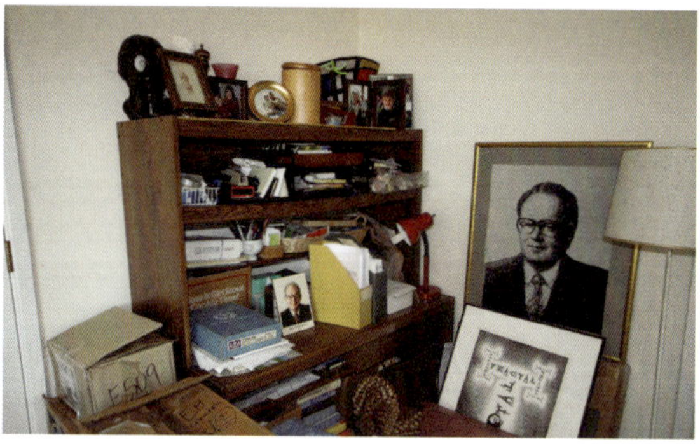

Fig 2. Benoît Mandelbrot's office, Cambridge, Massachusetts, 2011.

developed his concept of fractal geometry—and it is from that period that most of the work displayed in the exhibition at the Bard Graduate Center derives.

The vastness of his interests was echoed in the drawers overflowing with documents. It was pondered with astonishment by his contemporaries: "Benoît Mandelbrot is a continent," one exclaimed. "His book must have been written by thirty authors. It is just unbelievable how prolific this man was."[5] This very quality of his work became a curatorial challenge—and something of a treasure quest—while going through the overwhelming quantity of stuff left behind. In just three short visits, the whole "Mandelbrot continent" had to be explored and inspected closely; categories and criteria for selection had to be determined. The trick was to choose works that would best testify to Mandelbrot's image-centered working methods and, in a one-room gallery, serve to express the process of his scientific thinking in a way that would differ substantially from a written article. The firm decision to include one item would be undermined by the alluring content of the drawer next to it. Often, the size of an object would be diametrically opposed to its importance: Tiny graphical traces and doodled notes could turn out to be the most telling.[6] Sometimes, only parts of pictorial sequences were spotted, and the narrative stayed fragmentary. Planning this exhibition was a very intense and doubtlessly unforgettable experience.

Exhibiting: Shapes of Mental Processes

At the core of *The Islands of Benoît Mandelbrot* are numerous drawings, computer printouts, and other working materials found in the mathematician's office. They exemplify the practice of scientific visualization in a manner that has rarely, if ever, been exhibited before. In presenting these artifacts, the project takes seriously Bruno Latour's remark quoted in the epigraph: It is impossible to look inside scientists' heads, but the material residue of their experiments, their dealings with visible things, can be studied in detail. This project is based on the premise that working with material objects is an integral part of scientific thinking. If one looks at the "artifacts of science," one comes closer to obtaining a glimpse of scientific thinking than if one simply reads a scientist's publications, from which the traces of experimentation (both the true roads and the false leads) have generally been erased. Thinking does not take place in an ideal space devoid of objects, but through a use of materials that directly influences the shape and direction of a particular thought. *The Islands of Benoît Mandelbrot* aims to explore this dependence of scientific thought on material things and images. Therefore, the exhibition does not focus on the attractive illustrations scientists frequently produce for publication. Instead, it uses the detritus of scientific investigation to reveal the beauty of material thinking. This choice is reflected even in the title, which derives from a find in Mandelbrot's office, *The Island of Dr. Mandelbrot* (see fig. 1, Samuel essay), an early film depicting fractals. Mandelbrot's islands have a broad importance in his work and can be taken to symbolize the ambiguity of visual representation within scientific thinking, standing for two opposed powers of imagery—both the force of visual imagination as a driver of research and the dangers of believing that the mere appearance of a picture can in itself instantiate an unqualified truth. This seductive energy of images is linked to Mandelbrot's investigation of the tiny details on his computer pictures that he referred to as "islands" and whose mere appearance had to be questioned with regard to mathematical theory.

Pictures and pictorial technologies illuminate scientific questions because they can give a new form to the invisible and pose new questions for study. They can also have a retroactive influence on the subject as such, lending it permanent form.[7] In this exhibition, the role of pictures in shaping the way we think about science is not only examined on the basis of Mandelbrot's working materials; it also draws on the practices of Adrien Douady, an expert in the study of dynamical systems, and the early work of Otto E. Rössler, a prominent biochemist active in the related field of "chaos theory." Working materials from the mathematician and meteorologist Edward N. Lorenz, himself a founding father of "chaos theory," are also incorporated into the exhibition display thanks to the generous cooperation of the Library of Congress, which gave Bard access to sixteen unprocessed boxes of his notes, computer printouts, files, and photographs. One of the earliest and most important scientists to use weather simulations to study the chaotic behavior of dynamical systems, Lorenz made the study of chaos in the atmosphere a life-long occupation.[8] Browsing through his working materials, which have been preserved the way he left them upon his death in 2008, provided a perfect complement to visits to Mandelbrot's office some months before.

Mandelbrot's Work and the Picture Boom

Never have pictures been as widely used to produce and communicate scientific facts and fictions as in the last few decades.[9] At the same time, never has so much theoretical attention been paid to the role of picturing in the sciences as in recent years.[10] The most important reason for the proliferation of images since the 1980s is the fact that the computer can create visual representations. Yet digital images have been discussed primarily in relation to their "immaterial" properties or in contrast to their analog predecessors. The present project emphasizes visual continuities and discusses the materiality of digital forms.

Benoît Mandelbrot promoted the slogan "seeing is believing."[11] No other twentieth-century scientist's practice more thoroughly exemplifies the now proverbial "pictorialization of the sciences"[12] and the picture boom brought about by the computer. Mandelbrot was one of the earliest and best-known creators of digital pictures in a scientific and industrial context. Images of fractal geometry and the related field of chaos theory invaded various fields outside of science, rapidly becoming part of popular culture. The pandemic dissemination of this imagery represents a profound change in the role of visualization that continues to influence postmodern Western culture. Yet never before has an exhibition examined Mandelbrot's working material itself in relation to contemporary image theory.[13]

The Concept of the Catalogue

The essays in this catalogue all explore paths leading from the work of Benoît Mandelbrot to the role of material things in other sciences and disciplines. Mandelbrot's work lends itself singularly well to serving as a starting point in an investigation of how knowledge can be represented in pictures. The essays here spin a web of associations and observations, posing fundamental questions regarding pictures in science in general and illuminating multiple levels of visual thinking and image traditions using methods borrowed from art history.

The first text expands on the discovery of the "Island of Dr. Mandelbrot" and its implications. The notion of islands can be applied to various aspects of Mandelbrot's pictorial work, leading one to ask what role such concepts as imagination, visibility, and materiality can play in the scientific process. A prominent characteristic of pictures of chaos and of fractals is the fact that they appear as images generated or arranged in series. Accordingly, Matthias Bruhn explores the connection between chaos and its iconography and serial imagery. Wladimir Velminski presents a prehistory of representations of chance and statistics, one that Mandelbrot drew upon in order to produce images of natural structures. From a historical perspective, Margarete Pratschke discusses the theories developed by art historians to explain how visual order and disorder function as perceptual concepts. The computer has not only changed what we can see but also how we see. Mandelbrot's fractal geometry is a striking example of this process. With the aid of the computer, he discovered a new geometry and with it a new way of depicting and seeing nature. Jan von Brevern charts the extent to which this new view of nature differs from earlier models. Finally, Juliet Koss examines the tendency in Mandelbrot's work to transcend the distinction between the work of humans and that of nature—between *artificialia* and *naturalia*—as expressed by his claim that he made use of fractals as a tool for antimodernist art criticism.

Teaching Background

The project began with two concentrated courses. The first, "History and Style of Technical Images," took place in the fall of 2011. It was conceived as a broad survey of technical images over the centuries and in different media. It was initiated in collaboration with the Berlin-based research group Das Technische Bild and accommodated as visiting lecturers four of its active or associated members—Matthias Bruhn,

Margarete Pratschke, Jochen Hennig, and Karsten Heck. A highly motivated group of students and I conducted an exciting survey that ranged from sixteenth-century anatomical atlases to twenty-first-century nano images, exploring various visual strategies in science and technology. This was the first time that scientific images and artifacts were included as a distinct topic in the Bard Graduate Center's curriculum. In the second seminar, "Exhibiting Technical Images," a small group of students and I discussed how the finds from Mandelbrot's office might be displayed in the center's Focus Gallery and how this domain of private working materials could best be presented to the public. At nearly every class, we were able to sit around a table with some 300 images and objects spread out in front of us and to engage in lively discussions about the peculiarity of the objects and the best ways in which to display them. A unique highlight of the second semester was the filming of an interview with Mark R. Laff, who worked as a programmer for Mandelbrot at IBM between 1977 and 1979, and the production of a short film that has been integrated into the exhibition.

Acknowledgments

This project would not have been possible without the generous and patient cooperation of Aliette Mandelbrot, and I am sincerely grateful to her. I also wish to thank Leonard Bruno, Arnaud Chéritat, Régine Douady, Michael Frame, John Hubbard, Mark R. Laff, Gregory F. Lawler, Shaun Lovejoy, David Mumford, Alan Norton, and Otto E. Rössler, all of whom kindly provided information of a mathematical and factual nature during these past months of preparation. Particular thanks go to Nynne Christoffersen, Colin Fanning, and Andrew Goodhouse, my students in the second semester, for their creative engagement in our curatorial experiment, their fresh ideas, and adventurous enthusiasm in approaching an unconventional body of material that had not been used for an exhibition before. I am also indebted to the entire creative working group Das Technische Bild in Berlin for its active technical and personal support and most especially to the professional and dedicated faculty and staff of the Bard Graduate Center, which is led by Susan Weber: Ivan Gaskell, Laura Grey, Nina Stritzler-Levine, Ian Sullivan, Ann Marguerite Tartsinis, and Han Vu. I owe particular thanks to the dean, Peter N. Miller, who initiated this cooperative undertaking.

1 Bruno Latour, "Drawing Things Together," in *Representation in Scientific Practice*, ed. Michael Lynch and Steve Woolgar (Cambridge, MA: MIT Press, 1990), 21. I am grateful to Lena Bader for calling my attention to this passage.
2 Benoît Mandelbrot, interview with the author, Cambridge, MA, April 2008.
3 Gregory F. Lawler, telephone interview with the author, January 31, 2012.
4 Mandelbrot, quoted by A. Barcellos in "Interview of B. B. Mandelbrot," *Mathematical People* (Boston: Birkhäuser, 1984), 213. Mandelbrot held several positions as a visiting professor, and in 1999, he accepted a tenured position as Sterling Professor of Mathematical Sciences at Yale University. However, he felt generally indebted to IBM for giving him both the freedom and the technological support to extend his research interests beyond math ematical linguistics (the subject of his doctoral work) to the multiple fields that he ended up working in, which ranged from engineering to the mathematics of finance to the theory of physical turbulence; from hydrology to biology; from geomorphology to astro- and geophysics; and from physiology to risk management. Mandelbrot was born in Warsaw on November 20, 1924, and grew up in a Jewish Lithuanian family that immigrated to France. Before arriving at IBM in 1958, he worked for the French research organization Centre national de la recherche scientifique (CNRS). He died on October 14, 2010, in Cambridge.
5 Otto E. Rössler, interview with the author, Tübingen, Germany, June 1, 2007.
6 For a discussion of scribbles and laboratory notebooks in the history of science, see Hans-Jörg Rheinberger, "Scrips and Scribbles," *MLN* 118 (2003).
7 In terms of methodology, this approach follows that of the Berlin-based art-history working group Das Technische Bild, with whose collaboration this project was carried out. See Horst Bredekamp, Birgit Schneider, and Vera Dünkel, *Das Technische Bild. Kompendium zu einer Stilgeschichte wissenschaftlicher Bilder* (Berlin: Akademie, 2008); Matthias Bruhn and Vera Dünkel, "The Image as Cultural Technology," in *Visual Literacy*, ed. James Elkins, 165–78 (New York and London: Routledge, 2008). See also Bredekamp et al., "Bildwelten des Wissens," in *Bilder in Prozessen. Bildwelten des Wissens. Kunsthistorisches Jahrbuch für Bildkritik* 1 (Berlin: Akademie, 2003), and Bredekamp, "A Neglected Tradition? Art History as 'Bildwissenschaft.'" *Critical Inquiry* 29 (2003). Susan Dackerman refers specifically to the tradition of German Bildwissenschaft in her introduction to the catalogue of an exhibition of scientific images; see Susan Dackerman, ed., *Prints and the Pursuit of Knowledge in Early Modern Europe* (Cambridge, MA: Harvard Art Museums, 2011),19.
8 For his own account of his work see Edward N. Lorenz, *The Essence of Chaos* (Seattle: University of Washington Press, 1993). For his first seminal paper on chaos, see Lorenz, "Deterministic Nonperiodic Flow," in *Journal of the Atmospheric Sciences* 20 (1963).
9 See Martin Kemp, *Visualizations: The Nature Book of Art and Science* (Oxford: Oxford University Press, 2000) and, for a recent example of the importance of image technologies, see the case study on nanotechnology described in Jochen Hennig, *Bildpraxis: Visuelle Strategien in der frühen Nanotechnologie* (Bielefeld: Transcript, 2011). This development can be identified as a characteristic of the iconic turn; see Gottfried Boehm, "Die Wiederkehr der Bilder," in *Was ist ein Bild?* ed. Gottfried Boehm (Munich: Fink, 1994) and William J. T. Mitchell, "The Pictorial Turn," *Artforum* 30 (March, 1992).
10 For a survey of the related literature, see Hans-Jörg Rheinberger, Michael Hagner, and B. Wahrig-Schmidt, eds., *Räume des Wissens. Repräsentation, Codierung, Spur* (Berlin: Akademie Verlag, 1997); Bettina Heintz, Arnold Benz, and Jörg Huber, eds., *Mit dem Auge denken: Strategien der Sichtbarmachung in wissenschaftlichen und virtuellen Welten* (Vienna: Springer, 2001); David Gugerli and Barbara Orland, *Ganz normale Bilder. Historische Beiträge zur visuellen Herstellung von Selbstverständlichkeit* (Zürich: Chronos, 2002); Matina Hessler, ed., *Konstruierte Sichtbarkeiten. Wissenschafts- und Technikbilder seit der Frühen Neuzeit* (Munich: W. Fink, 2006).
11 Benoît Mandelbrot, *The Fractal Geometry of Nature* (New York: Freeman, 1982), 21. The biblical origin of this punchy formula is related as much to touching as to seeing; see Glenn W. Most, *Doubting Thomas* (Cambridge, MA: Harvard University Press, 2005). See also Hans Belting, *Bild und Kult: eine Geschichte des Bildes vor dem Zeitalter der Kunst* (Munich: Beck, 2000) and Christof L. Diedrichs, *Vom Glauben zum Sehen: Die Sichtbarkeit der Reliquie im Reliquiar: Ein Beitrag zur Geschichte des Sehens* (Berlin: Weissensee, 2001). It has also been used as a title of an exhibition at the New York Public Library that surveyed 700 years of the history of scientific and medical illustrations. See Jennifer B. Lee and Miriam Mandelbaum, *Seeing Is Believing: 700 Years of Scientific and Medical Illustration* (New York: New York Public Library, 1999).
12 Heintz, Benz, and Huber, eds., *Mit dem Auge denken*, 9.
13 The project is based on the findings in my dissertation, "Die Form des Chaos," Humboldt-Universität zu Berlin, July 2011.

The Visibility of Islands: On Imagination, Seduction, and Materiality

Nina Samuel

Prologue: Chasing Traces; or, The Blind Spot of a Discussion

One day in the late 1980s the mathematician Alain Connes and the neurobiologist Jean-Pierre Changeux sat down together in Paris to hold "conversations on mind, matter, and mathematics." Despite being eminent men of knowledge, they were utterly unable to come to an agreement.[1] The two adversaries were discussing a fundamental dilemma, the ramifications of which extend far beyond the discipline of mathematics: Can the world of thinking be separated from the world of things? And if not, how can we conceive of the interactions between the two realms? Does thinking—together with mathematical objects—have a material foundation? Does it leave traces? And are thoughts dependent on the instruments that make them possible and give them their shape? Connes argued that there was a mathematical world "independent of us, which escapes sensory apprehension." Changeux's position was the opposite: "Granting your position for the sake of argument, I've tried to see where this world could be, to see what *evidence* of it there is in nature. If you form the hypothesis that this mathematical world exists outside of us, and if you call yourself a materialist, then it seems to me you're obliged to give it a material basis."[2] The men could not agree.

What is striking about this conversation, however, is not that the two men would be in the grip of diametrically opposed ideas. After all, for centuries people have been disputing such things. No, what was striking was what they left out: Never once was the topic of mathematical *images* raised in any detail—and it was precisely this topic that might have shown them the way to mutually acceptable answers. In this essay, I will attempt to draw attention to the ways in which images and objects have either triggered mathematical thinking or served as its subject matter and helped to shape it.

In what follows, I will touch on three moments in the "personal-pictures" history of fractals and chaos studies as modes of visual thinking, which is being presented here for the first time. The analysis of the private working materials of Benoît Mandelbrot stands at the beginning of this essay, taking up the greatest part, to be followed by two briefer discussions of scientists from the related fields of dynamical systems and chaos theory, Adrien Douady and Otto E. Rössler. Along the way, the surprisingly complicated relationship between mathematics and its images and representations will be explored, along with two different meanings of "mathematical islands," one figurative and one abstract, each pointing to an antipodal power of visual representation in the thinking process. Pictures can release a productive, imaginative power that advances thinking, but attached to them is a seductive and potentially ungovernable energy as well, one that causes frictions between vision and thought and even, potentially, egregious errors. These two sides of pictures are inseparable and equally necessary.

If all three scholars used images in fundamentally different ways, and if the juxtaposition of their different methods of deriving mathematical ideas from the world of the visible and the material will help distinguish three basic different modes of visual thinking, it is equally true that the thinking of

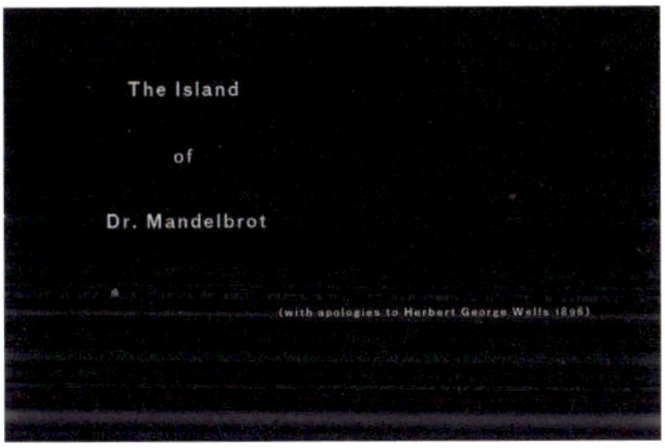

Fig. 1. Benoît Mandelbrot. *The Island of Dr. Mandelbrot*, ca. 1971–75. Computer-generated movie. Still. Collection Aliette Mandelbrot.

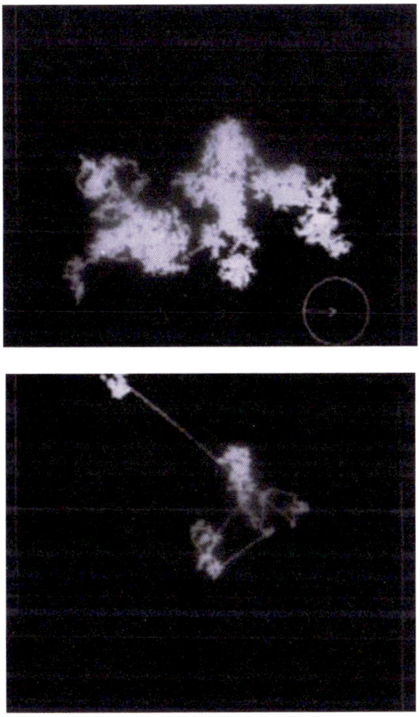

Fig. 2. Benoît Mandelbrot. *Paul Lévy Motion*, ca. 1971–75. Computer-generated movie. Hirsh Lewitan, programmer. Stills. Collection Aliette Mandelbrot.

all of them was deeply grounded in visual representations and that their work would not have been conceivable without images. For all of them, experiments with early computer visualizations proved decisive, though hand drawing and other modes of visual and material manipulation were equally indispensable as modes of thinking and understanding in the realm of mathematical ideas.

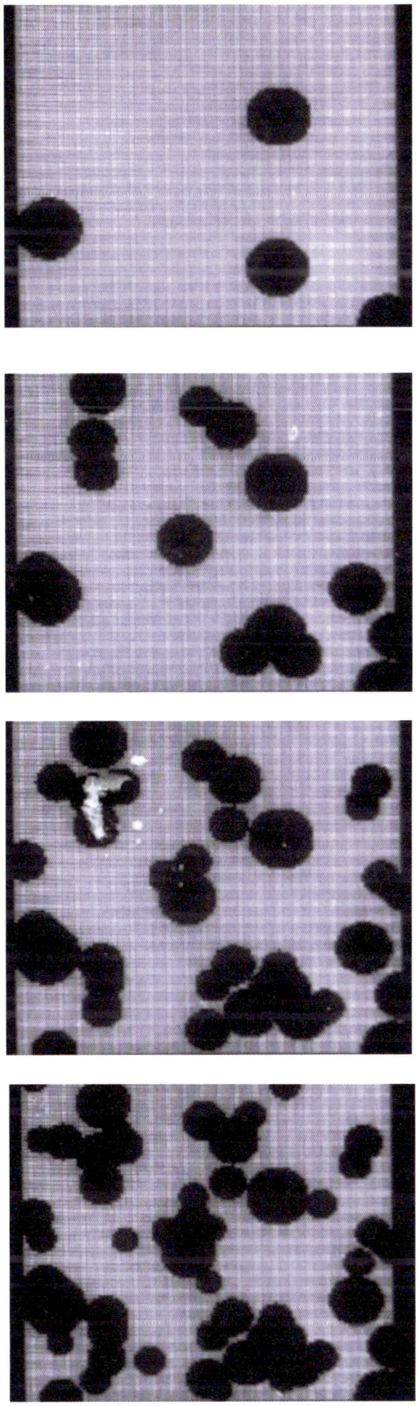

Fig. 3. Benoît Mandelbrot. *Circular Cutouts in Two Dimensions*, ca. 1971–75. Computer-generated movie. Stills. Collection Aliette Mandelbrot.

The Discovery of the Island: Playful Sequences of Similarities

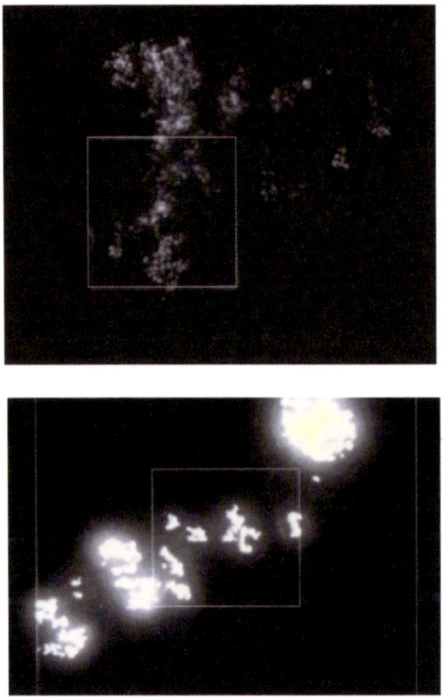

Fig. 4. Benoît Mandelbrot. Zooms into *Paul Lévy Motion*. ca. 1971-75. Computer-generated movie. Stills. Collection Aliette Mandelbrot.

When seeking new insights, I look, look, look, and play with many pictures. (One picture is never enough!)[3]
—Benoît Mandelbrot

Viewers fortunate enough to view Benoît Mandelbrot's first fractal computer animations, made in the early 1970s, may well experience a jolt of surprise at the sudden confluence of the worlds of science and science fiction when the words "The Island of Dr. Mandelbrot" appear on the screen for a few seconds (figs. 1–5).[4] Produced by Mandelbrot and programmer Hirsh Lewitan at IBM on a Stromberg Datagraphix 4020, the animations feature the jagged objects that Mandelbrot began referring to in 1975 as "fractals": collections of dots, lines, surfaces, or solids, "loosely characterized as being violently convoluted and broken up,"[5] that cannot be comprehended by means of Euclidean geometry. The films depict such structures as animations generated by chance algorithms. One sees three-dimensional compositions—dots, lines, the relief structures produced somewhat later (his so-called "imaginary continents")—slowly rotating in front of a black background.

In the line below the intertitle "The Island of Dr. Mandelbrot" (fig. 1) is proof that Mandelbrot himself associated his fractals with the 1896 fantasy novel *The Island of Dr. Moreau*. There, he added in tiny, barely legible letters, "(with apologies to H. G. Wells)" (fig. 6).[6] With this twist on the title and the mock "apology," Mandelbrot deliberately established a connection between "science and science fiction" and, with gentle, endearing irony, related his own work to that of a long line of scientists seen as godlike, if weird, creator figures.[7] There is a hint in these first animated simulations of a classic dream of the early pioneers of computer graphics, namely that they might, with the aid of the computer, create a new world, a second reality. By way of the film's montage technique, that world was given a quite specific structure, one expressed in repeated,

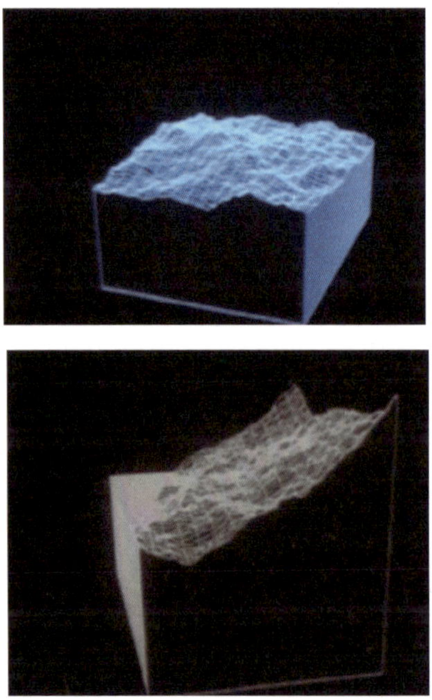

Fig. 5. Benoît Mandelbrot. *Imaginary Continents*, ca. 1971-75. Computer-generated movie. Stills. Collection Aliette Mandelbrot.

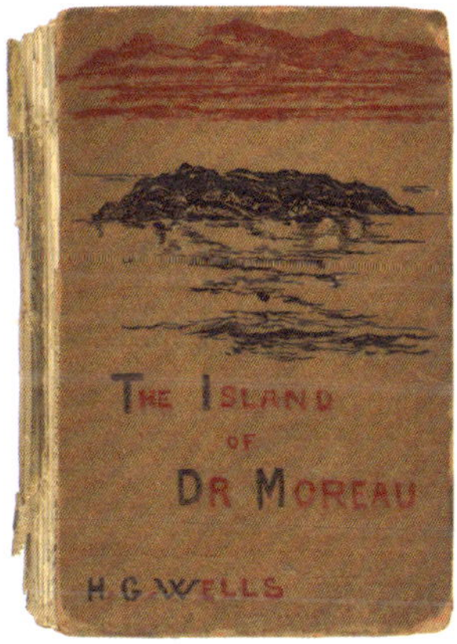

Fig. 6. H. G. Wells. *The Island of Dr. Moreau* (London: William Heinemann, 1896). General Research Division, The New York Public Library, Astor, Lenox and Tilden Foundations.

rapid zooms (see fig. 4). The viewer is drawn in closer and closer—without ever landing—as greater and greater enlargements appear. Zooms first entered film history as a dramatic effect in Hitchcock's *Vertigo* (1958), and ten years later they were employed by Charles and Ray Eames in the first version of their famous film *Powers of 10* (*Rough Sketch*, 1968) as a way of staging a journey between atoms and galaxies (fig. 7).[8] Mandelbrot knew the Eameses, who also worked in association with IBM, but he characterized their work as "violently antifractal propaganda."[9] Unlike the designers' films, Mandelbrot's did not emphasize the differences between levels of magnification;

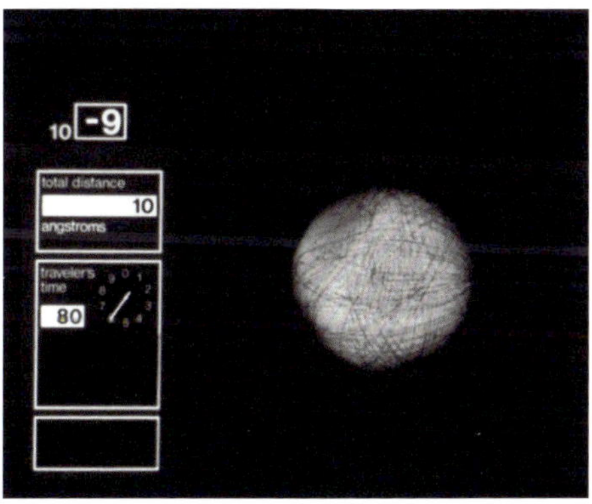

Fig. 7. Charles and Ray Eames. *Rough Sketch*, 1968. Still. © Eames Office, LLC.

instead, as he was at pains to point out, they created a cosmos out of similarities. The sections marked with squares in the film revealed in increasing close-up the same frayed structures as could be seen from farther out. In this way, Mandelbrot dramatically illustrated one of the most important geometric characteristics of fractals—their self-similarity.

Perhaps the two key features of Mandelbrot's picture making—as well as that of other scientists working in the field—were his attention to similarities and the works' appearance in a series. Mandelbrot routinely worked with sequences of pictures; virtually every shape appeared again and again, with the very slight differences between them coming only in the form of minimal adjustments to definition, slight shifts in picture details, tiny modifications of parameters, et cetera. There were piles of variant picture shapes that were so similar as to be virtually indistinguishable. That this is characteristic of pictures from this field of study raises the question of how picture sequences and single pictures relate to one another. Can our observation of the material evidence of the working process tell us anything about how serial pictures facilitate mathematical understanding? The pictorial work of Benoît Mandelbrot provides the first answer to this question.

In the Beginning Was the Image; or, In Praise of a Medium Without Touching It

To see is to believe.[10]
—Benoît Mandelbrot

In 1985, when Columbia University awarded Benoît Mandelbrot the Barnard Medal for Meritorious Service to Science, it did so with the following justification: "In the great tradition of natural philosophers past you looked at the world around you on a broader canvas."[11] Within roughly ten years after his first computer animations, Mandelbrot had managed to create a geometry that brought him renown far beyond the boundaries of mathematics—a geometry associated with a specific worldview. How did he himself view the world? Or perhaps better: How did the public in 1985 view Mandelbrot's view of the world?

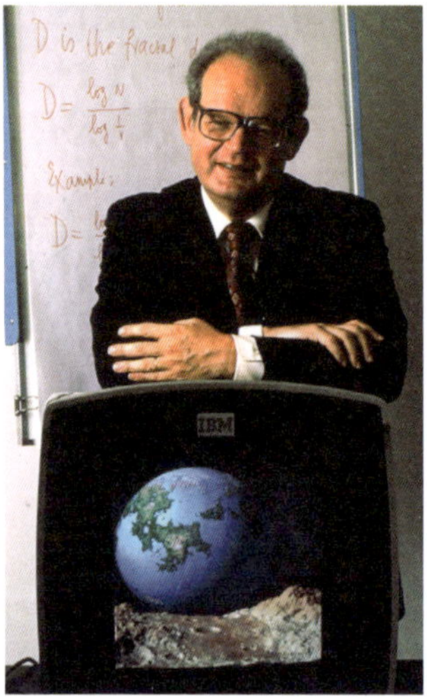

Fig. 8. Benoît Mandelbrot with computer-screen display "Planetrise over Labelgraph Hill (Souvenir from a Space Mission that Never Was)," programmed by Richard Voss, 1982. Press photograph, ca. 1985. Image originally created by IBM Corporation.

In a press photograph published by IBM around the time (fig. 8), a smiling Mandelbrot can be seen leaning on one of the firm's computers.[12] Behind him, there is a large paper board on which is written the formula for a general definition of the fractal dimension (D). (This dimension is an index for the varying degree of roughness of fractal structures. It quantifies its complexity as a ratio of the change in detail to the change in scale. The dimension D is always a decimal number and distinguishes fractals from Euclidean geometry, in which there are only whole-number dimensions for lines, planes, and solids.[13]) The word "example" also appears on the board, but the rest of the line is obscured by Mandelbrot's right shoulder, and thus the role of the example is taken up by a seemingly cosmic scenario on the computer monitor on which Mandelbrot rests. Programmed by his colleague and friend Richard Voss and bearing the title "Planetrise over Labelgraph Hill (Souvenir from a Space Mission that Never Was),"[14] the image presents a sphere that looks much like Earth—its colors had been manipulated to create this effect—looming up out of the darkness of the monitor above a jagged and shadowed hilly landscape covered with round craters. It is one of the highlights of the use of fractals for pictorial fantasizing for which Mandelbrot became known to the public. Shifting the formula literally into the background, it not only presents the picture the formula produces, it illustrates the prominence of images in his way of thinking and working—specifically digitally produced images. Indeed, the decision by the IBM photographer to stage this photo as he did makes Mandelbrot's earlier, quite non-public reference to his work as an imaginary "Island of Dr. Mandelbrot" seem almost visionary. Evidently, the dream of being able to create a second nature with the aid of fractals, and thereby to have deciphered mathematically the principle behind the creation of the world, was now Technicolor reality.

The view of the world that Mandelbrot presented on his monitor had unmistakable precursors in pre-digital picture history. Not without reason does it recall scenes from one of the first American science-fiction classics, *Destination Moon* (1950) (fig. 9). And even though Mandelbrot gave out that the IBM photo showed a "space mission that never was," such flights were indeed taking place quite frequently then—in movie theaters. Immediately after the publication of Mandelbrot's first English-language book on fractals, filmmakers began usurping the

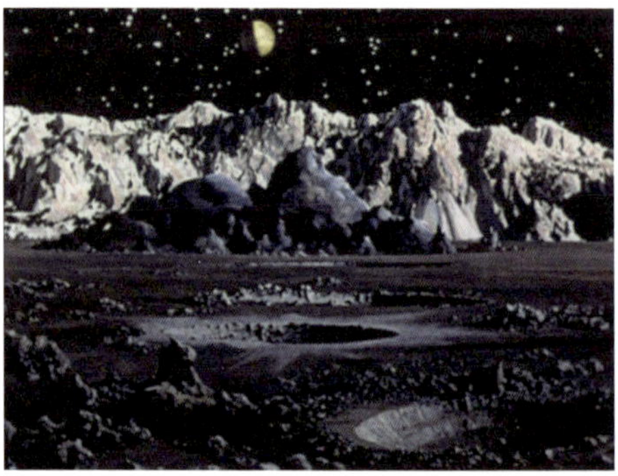

Fig. 9. *Destination Moon*, 1950. Still. Corinth Films.

fractal simulation of nature. It was not long before the first planetary scenes wholly produced by means of his fractals appeared, in *Star Trek II: The Wrath of Khan* (fig. 10).[15]

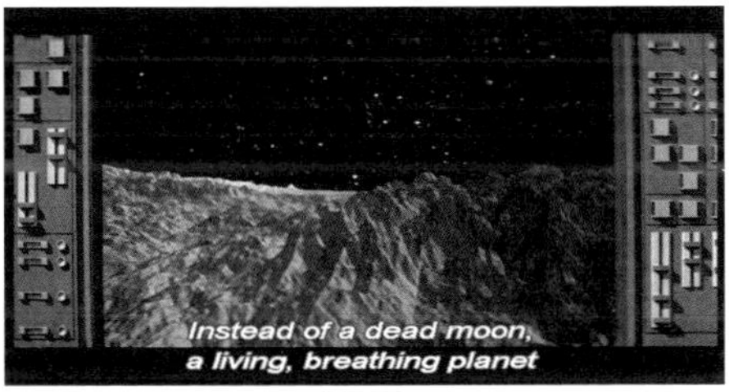

Fig. 10. *Star Trek II: Wrath of Khan.* Still. ©1982 Paramount Pictures. All Rights Reserved.

Mandelbrot's fondness for algorithmically simulated cosmogonies with a science-fiction aesthetic, along with his frequent use of the provocative phrase "to see is to believe," greatly influenced the way his work was perceived.[16] In later statements, he appeared to glorify visual representation. In a subtle attack on the Old Testament proscription against graven images, he rephrased the opening line of the Gospel of St. John: "Wouldn't 'In the beginning was the word' have to be replaced with 'In the beginning was the picture?'"[17] Mandelbrot's unabashed devotion to pictures was not universally shared; it occasioned heated criticism both within mathematics and outside it.[18] In their eternal repetition and similarity, images of fractals and chaos were seen by art historians as breaking with classical mimesis theory, as threatening to collapse "the difference between picture and reality itself."[19] It is significant, however, that these appraisals were primarily based on popularized images that were not produced in the process of mathematical research. They were also reactions to the influence the reception of such pictures was having on postmodern Western discourse: Fractals became a flagship of simulation theory and a symbol of "new" or even "immaterial" picture mediums that were categorically difficult to understand and seemed to break with all pictorial tradition.[20]

Given his close identification with the computer, it may seem paradoxical that Mandelbrot was wholly incapable of programming one himself. All his computer graphics were team efforts. Not only did he gather about him a host of programmers, he also seems to have been conflicted in his relationship to the device. All his manuscripts and text corrections were either written by hand or typewritten (see fig. 1, page 154). He avoided touching a computer himself—an evasion that was well known in the firm, and that led the IBM photographer to jokingly comment that he had obviously done so in the photo (see fig. 8). In good humor, Mandelbrot corrected him, claiming that he had leaned against the monitor, to be sure, but had not "touched it *with his hands*."[21] The anecdote may only be embroidered oral history, but it touches upon something fundamental: There is a significant discrepancy between the way scientists employ images and the way those images are seen by the public. As the present exhibition aims to show, in cases where discussions are based on popularly known imagery, as was long the case for fractals and chaos studies, this gap can be immense. Indeed, it provides both material and motivation for the reevaluation of the role of computers as interfaces between the world of thinking and the world of things. On the other hand, the surprising appearance of literary references in Mandelbrot's newly-discovered early films (see figs. 1–5) indicates a second fundamental and opposing dynamic: It was not only in the popularization of such images that they came to be laden with meanings that transcended mathematical or physical issues. Playing with images—with sequences of similarities—has to be understood as a genuine part of research and thinking in general.

Mandelbrot's focus on visual representation raises questions that can only be dealt with after more detailed analysis of both his work as a whole and his use of images in particular from the perspective of image theory and epistemology. His equation of seeing and believing is doubtless unproven, equally indefensible whether one dismisses it as evidence of a naive faith in images or takes it as either a rhetorical flourish or a deliberate, tongue-in-cheek provocation. It still stands as a polarizing, rhetorically pointed reaction against what he saw as the devastating effects of certain late-nineteenth-century developments in mathematics.

Mandelbrot was firmly convinced, and never grew tired of emphasizing, that

> the use of computer graphics is now in the process of altogether changing the role of the eye. The hard theoretical sciences had banished the eye for a long time, and many observers used to believe, and even hope, that it would remain banished forever. But computer graphics is bringing it back as an integral part of the very process of thinking, search, and discovery.[22]

Where had "the eye" gone, and how had it disappeared?

The Picture as Menace

I see it, but I don't believe it.[23]
—Georg Cantor

The progress of mathematics is not always smooth. At the end of the nineteenth century, for example, the French mathematician Charles Hermite reported to his colleague Thomas Jean Stieltjes that calculus was afflicted with a "curse" and that he had turned away in fear and horror "from the lamentable plague of functions without derivatives."[24] Mathematics had been in the midst of a crisis for some time, one brought about by deviations from classical assumptions about elementary concepts.[25] At the core of the crisis were functions whose values sprang endlessly back and forth between two constants and for that reason could not be precisely drawn. Such imprecise constructions had previously been banished from mathematical discourse, but even so, they now threatened to lead "to a frontier of the intellect."[26]

These oscillating, interdimensional structures could have been a further reason why in the mid-1970s—at the very moment he was coining the word "fractal"—Mandelbrot chose to make the allusion to H. G. Wells's fantasy novel: The fictional Dr. Moreau had tried to create a new world with man-animal hybrids, and it was for a similar purpose that Dr. Mandelbrot was using those mathematical hybrids from the nineteenth century. Henri Poincaré, among others, had referred to these new functions as "monsters" of mathematical logic and labeled them outrageous attacks on the intelligence of his precursors.[27] Roughly a hundred years later, Mandelbrot touted these so-called "monsters" as the first mathematical fractals. He liked to repeat this blasphemy as a way of pointing to his own achievement in having made out of "monsters" a "geometry of nature" that, above all, have "the advantage of being beautiful."[28] By repeatedly emphasizing that he had incorporated monsters into his new, digital nature, he was also aligning himself with classical sixteenth-century natural history, in which nature itself became the realm of "exceptions, the rare, and the marvelous."[29]

The crisis in mathematics was also one of seeing. When the mathematician Georg Cantor questioned the principles behind the Euclidean idea of dimension, he rightly associated the shock his discoveries had given him with a disbelief in the reliability of his vision: "I see it, but I don't believe it."[30] It was as an ironic twist on Cantor's comment that Mandelbrot chose his own motto: "to see is to believe." Indeed, Mandelbrot accused the mathematicians of the turn of the twentieth century of "blindness" for failing to recognize the visual qualities and the clear import of their phenomena.[31] At the heart of the crisis was a growing mistrust of the cognitive ability of the eye and accordingly of pictorial representations. The more radically even elementary fields of mathematics broke away from the way humans conceive of space visually, the more difficult it became for the eye to follow the discipline, and the more inadequate and misleading even schematic graphics appeared to be. This "plunge of mathematics into the unimaginable" in response to new "monstrously oscillating" functions has been referred to as "iconoclasm" by the media historian Bernhard Siegert.[32]

And yet there are a few surviving visualizations of the first mathematical fractals: simple line drawings, highly schematic and directly related to structural steps described in texts.[33] They record the recognition that the human eye could no longer follow geometry's experimental thinking since only a few structural steps led it below the visibility threshold. These drawings were symptomatic of a historic change in mathematics, one in which the relationship between geometry and the world of experience, between thinking and seeing, was renegotiated. On the one hand, what made it possible for the "monsters" to become the subject of official mathematical discussion was that spatial perceivability (*Anschauung*) had ceased to be a normative principle at the time.[34] Given that at least a few mathematicians recorded their deliberations in graphic form, it is clear that a certain value was still accorded to the picture and to seeing. But generally it was clear that in order for constructions that broke so radically with fundamental assumptions to be integrated into mathematics, the discipline itself had to be redefined.

The crisis was soon perceived as a threat that shook mathematics to the core; in 1921 Hermann Weyl described it as a "*Grundlagenkrise*" (foundational crisis) that threatened to destroy the "inner solidity and certitude"[35] of analysis. A way out of the crisis was finally proposed by David Hilbert with his redefinition of the "foundations of geometry."[36] As early as 1899, he showed that the discipline could be axiomatized, an initial but momentous step toward the accommodation of the uncertainties caused by an explicit rejection of visualization.[37] Hilbert had concluded that geometrical figures were unworthy of the mathematician's "superior intellect" and as a "practical constraint" were contrary to "freedom."[38] With Hilbert, mathematical symbols became autarkical. They no longer meant or referred to anything outside of mathematics.[39] By the end of the 1920s, the prevailing view was that "objects of present-day mathematics . . . are no longer tangible things; . . . their structures . . . are of the most extreme abstraction. . . . [There is a] tendency toward intellectualization, a disassociation from everything material."[40]

In such a climate, the ground was prepared for the ascendency of a group of French mathematicians who would concentrate and intensify the iconoclastic tendencies in mathematics. They radicalized Hilbert's formalistic approaches and considered pictures to be not only problematic or inadequate but a menace to mathematical thinking. Beginning in 1935, they grouped together under the name "Nicolas Bourbaki." The Greek-sounding name, which was derived from that of a general in the Franco-Prussian War, served them as a collective pseudonym, the reasons for which they never fully revealed.[41] They were united in their championing of a method characterized by its rigorous structuralism and formalism: For them, structure as a "treasury of abstract forms"[42] was to be the true subject matter and medium of mathematics. It would be of crystalline design, "transparent down to the depths."[43] With the publication in 1939 of the first volume of their *Éléments de Mathématique*, they became one of the most influential authorities in mathematical research and teaching.

All of their authorized publications were similar in one crucial respect: There were no illustrations. According to the Bourbakist Pierre Cartier, their rejection of pictures was true to "the Euclidean tradition," which in their view was originally non-pictorial. In addition, he stated, most of the members were puritanical, among them a great many Protestants and Jews—"Old Testament people"—and as such they were ill-disposed to visual representations of religious and mathematical truths.[44] Their iconoclasm was unwritten but extraordinarily effective. For example, the iconoclastic imperative is implicit in the suggestion that mathematical intuition is of importance but in a "manner wholly unrelated to ordinary sense perceptions."[45] Because mathematics was to be thought of as a "poetry of ideas,"[46] the "original visual content" of mathematical figures needed to be "consciously excluded."[47] In line with this call for precision, the discipline was to be entirely transformed into text.

Linked to this radical rejection of images was a correspondingly forceful admission of their power. This was expressed most blatantly by a pictorial symbol shaped like a reversed "S" that the Bourbakists invented for their publications (figs. 11 and 12). In the "directions for use" (*Mode d'emploi de ce traité*) that preceded their volumes, they explained its meaning and called it a *tournant dangereux* (dangerous turn).[48]

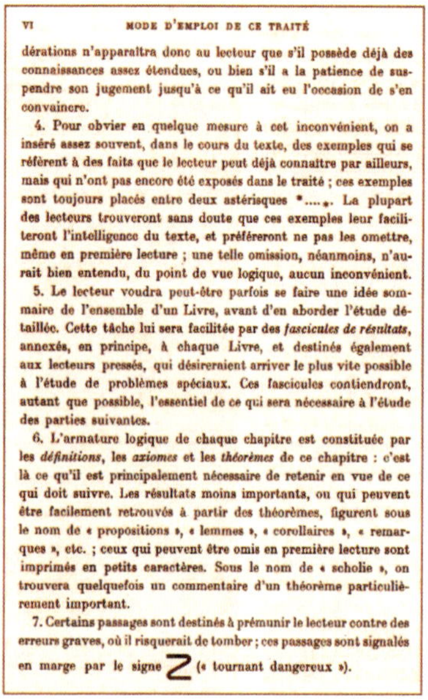
VI MODE D'EMPLOI DE CE TRAITÉ

dérations n'apparaîtra donc au lecteur que s'il possède déjà des connaissances assez étendues, ou bien s'il a la patience de suspendre son jugement jusqu'à ce qu'il ait eu l'occasion de s'en convaincre.

4. Pour obvier en quelque mesure à cet inconvénient, on a inséré assez souvent, dans le cours du texte, des exemples qui se réfèrent à des faits que le lecteur peut déjà connaître par ailleurs, mais qui n'ont pas encore été exposés dans le traité ; ces exemples sont toujours placés entre deux astérisques *.... *. La plupart des lecteurs trouveront sans doute que ces exemples leur faciliteront l'intelligence du texte, et préféreront ne pas les omettre, même en première lecture ; une telle omission, néanmoins, n'aurait bien entendu, du point de vue logique, aucun inconvénient.

5. Le lecteur voudra peut-être parfois se faire une idée sommaire de l'ensemble d'un Livre, avant d'en aborder l'étude détaillée. Cette tâche lui sera facilitée par des *fascicules de résultats*, annexés, en principe, à chaque Livre, et destinés également aux lecteurs pressés, qui désireraient arriver le plus vite possible à l'étude de problèmes spéciaux. Ces fascicules contiendront, autant que possible, l'essentiel de ce qui sera nécessaire à l'étude des parties suivantes.

6. L'armature logique de chaque chapitre est constituée par les *définitions*, les *axiomes* et les *théorèmes* de ce chapitre : c'est là ce qu'il est principalement nécessaire de retenir en vue de ce qui doit suivre. Les résultats moins importants, ou qui peuvent être facilement retrouvés à partir des théorèmes, figurent sous le nom de « propositions », « lemmes », « corollaires », « remarques », etc. ; ceux qui peuvent être omis en première lecture sont imprimés en petits caractères. Sous le nom de « scholie », on trouvera quelquefois un commentaire d'un théorème particulièrement important.

7. Certains passages sont destinés à prémunir le lecteur contre des erreurs graves, où il risquerait de tomber ; ces passages sont signalés en marge par le signe Z (« tournant dangereux »).

Fig. 11. Nicolas Bourbaki. Page from *Éléments de Mathématique* (Berlin: Springer, 2007). With kind permission of Springer Science and Business Media.

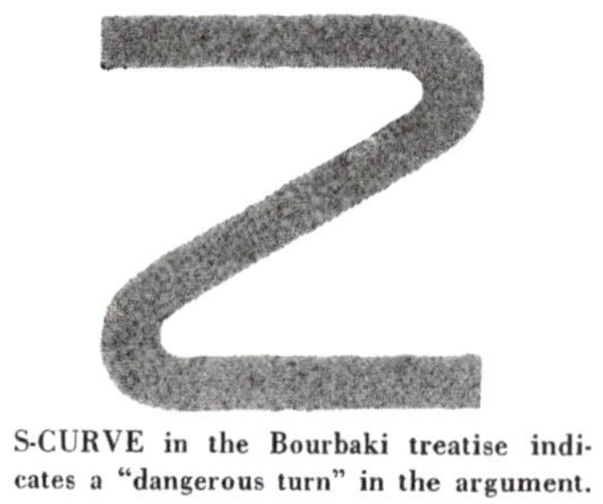

Fig. 12. Paul Halmos. "The 'S-Curve of the Bourbakists." From "Nicolas Bourbaki," *Scientific American* 196 (1957): 93. Reproduced with permission. Copyright © 2012 Scientific American, Inc. All rights reserved.

Whenever their reasoning threatened to become precarious, they presented the reader with this hair-pin bend: Just by glancing at the page, the reader would know to prepare for intellectual hardship or logical shoals. The way the bold, squat typography of the sign visually cuts through the arrangement of the typesetting demonstrates that the "iconoclastic" mathematicians conceived of their publications' layout as a pictorial element in its own right. By understanding and using the image qualities of the typeface, they were able to convey the importance of the symbol in an intuitive and immediate way.

Digging deeper into the pictorial history of the sign the Bourbakists chose, an even more striking level of meaning surfaces. On a formal level, the mathematicians had resorted to the centuries-old *figura serpentinata* as a primal expression of the creative energy in nature and in art, one that can be identified as a universal "symbol of all drawings" and "the sum of the visual."[49] In 1753, the English artist William S. Hogarth defined it in *The Analysis of Beauty* as an emblem of "Variety" and accordingly a symbol of all forms of pictorial expression (fig. 13).[50] Some two hundred years after that, its rich meanings were still being celebrated and investigated by artists such as Paul Klee,

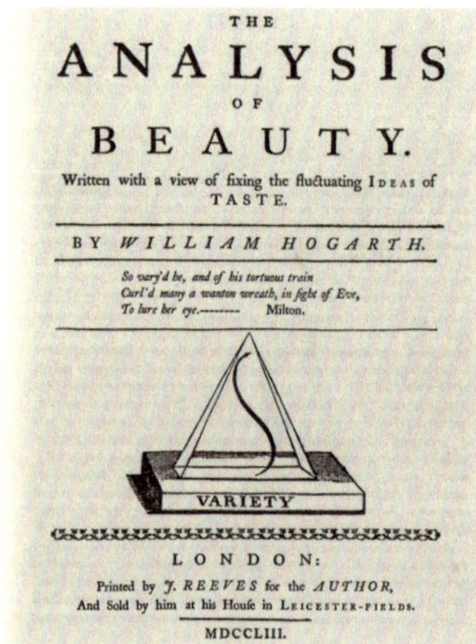
THE

ANALYSIS

OF

BEAUTY.

Written with a view of fixing the fluctuating IDEAS of TASTE.

BY *WILLIAM HOGARTH.*

So vary'd he, and of his tortuous train
Curl'd many a wanton wreath, in sight of Eve,
To lure her eye.-------- Milton.

LONDON:

Printed by *J. REEVES* for the *AUTHOR*,
And Sold by him at his House in LEICESTER-FIELDS.

MDCCLIII.

Fig. 13. William Hogarth. Title page from *The Analysis of Beauty* (London: J. Reeeves, 1753).

whose graphic work explores the active dynamics of convoluted lines as a foundation for both artistic practice and image-theoretical inquiry (fig. 14).[51] The Bourbakists' stylized serpentine line was used to flag spots that spelled the same kind of danger for the "realm of pure thought"[52] that "imprecise" pictures in general did. Thus they banished what they

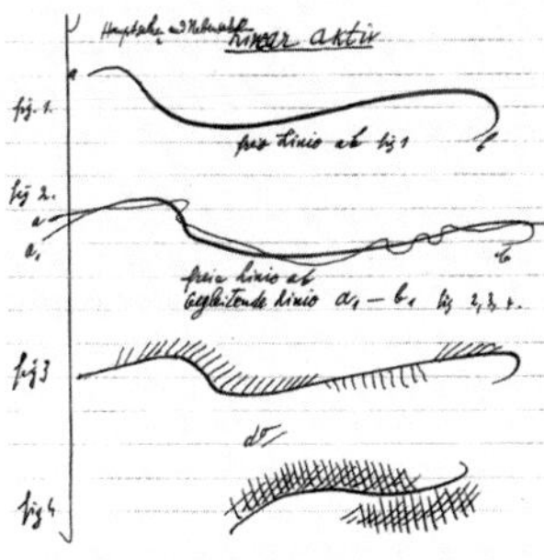

Fig. 14. Paul Klee. "Examples for Active Lines" from *Beiträge zur bildnerischen Formlehre*, November 14, 1921, fol. 8, fig. 1–4. Art and Architecture Collection, Miriam and Ira D. Wallach Division of Art, Prints and Photographs, The New York Public Library. Astor, Lenox and Tilden Foundations.

viewed as the dreaded destructive power of the visual by means of the visual. The Bourbakist S-curve became a kind of hex sign that called upon the magical power of the pictorial in a way reminiscent of the ancient stylized appropriation of snakes as a cultic means of defending against them.[53] One may assume that they were unaware of the traditional iconic meaning of the sign, but even so, it functioned as a thorn in the flesh of pure, formalized doctrine.

In the following decades, formalistic representation was pitted against geometrical, pictorial representation in official mathematical discourse. The dogma of anti-pictorialism left drastic traces in the lives of certain mathematicians, but it was becoming apparent that pictures could not be completely banned. In one example, the mathematician and Fields Medal–winner Alexander Grothendieck first made a name for himself with works in the Bourbakist mode. But just before his retirement, he experienced a "pictorial turn." He not only invented his own "picture theory of mathematics," in the course of this change he literally condemned formalism to hell as being inimical to life.[54] In 1971, after one of his last official lectures at the University of Bielefeld, in Germany, he left a drawing in the seminar logbook in which a mathematical notation introduced by the Bourbakists is consumed in raging flames and surrounded by demons armed with pitchforks (fig. 15). The drawing was embedded in the following commentary:

> Witch's Kitchen 1971. Riemann-Roch Theorem: The "dernier cri": The diagram [displayed] is commutative! To give an approximate sense to the statement about f : X → Y, I had to abuse the listeners' patience for almost two hours. Black on white (in Springer Lecture Notes) it probably takes about 400, 500 pages. A gripping example of how our thirst for knowledge and discovery indulges itself more and more in a logical delirium far removed from life, while life itself is going to Hell in a thousand ways—and is under the threat of final extermination. High time to change our course!

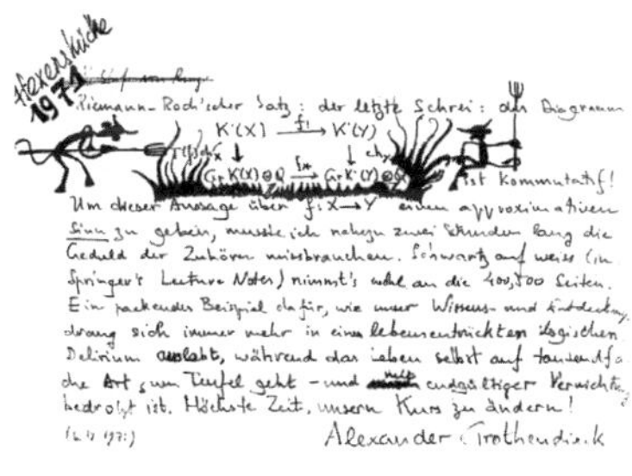

Fig. 15. Alexander Grothendieck. Entry in the colloquium book from Bielefeld University, 1971. Pen on paper. Reproduced with the kind permission of the Department of Mathematics of the University of Bielefeld from the department's 1971 colloquium book.

That change of course seemed to be promised by the invention of the computer as an image-making machine. In an ironic turn, the picture-rejecting method based on Hilbert and elevated to the status of dogma by the Bourbakists had meanwhile become the theoretical point of origin for the computer. The language of computer programming was derived from a strictly textual mathematics, yet with the blossoming of that language in computers it became possible to convert structures no longer accessible to three-dimensional representation into a new form of visibility. The capacity of the computer to continuously repeat a calculation and with each iteration add further detail to the graphic facilitated the generation of this new kind of imagery, which before had been seen as betraying human intuition. Because fractal structures are composed of an infinite number of details, the new technology seemed especially suited to making such highly complex shapes graspable. Technically speaking, fractals and chaotic dynamics lack closed analytic solutions, which means one has to actually count through all the states produced by the dynamics. A human being is at a loss here, particularly because even the calculation of a single step is a challenge. However, the computer is able to mimic the generative process of the dynamics in hand and consequently to visualize patterns inherent to the equation that could not be seen before. In other words, the dynamics themselves are generatively produced, much like a dynamical process in nature.

Mandelbrot held the Bourbakists in part responsible for his having turned his back on Europe and subsequently on academia. With the help of the computer, proof would be provided that the rejection of imagery had been a mistake. It was with a heartfelt sense of triumph that toward the end of his life Mandelbrot was able to declare that "Bourbaki is over."[55]

Spotting Islands, Cartographies of Chance: Analogy as a Principle of Thought

I see things everywhere.[56]
The eye is not specialized.[57]
—Benoît Mandelbrot

Mandelbrot once characterized the aim of his scientific life's work as the restoration of a "concrete meaning" of sight in which the seeing eye is accorded the status of a scientific instrument.[58] The reason it was such an effective instrument, he believed, was that it functioned in an interdisciplinary way. To Mandelbrot, this was owing to one of the main epistemic merits of seeing: "The problem with calculations is that they become very much specialized in the different fields. On the contrary, the eye is not specialized. It is a universal tool."[59] That each academic discipline practices its own calculation method and relies on specific knowledge acquired over centuries often renders it barely accessible even to neighboring disciplines. This could be alleviated, he insisted, through the cultivation of seeing, a theory he was fond of citing retrospectively to explain his own interdisciplinary approach and the universal validity of his fractal theory: They were, he claimed, achievements of the "unspecialized eye." To him, formulas and pictures were inseparable: "[E]very formula evoked in my mind a completely spontaneous shape."[60] But this did not imply that the eye of a "layman" was what was required; rather, he was claiming that seeing had no need of specialization to function as an instrument of understanding.

But this raises the question of how "unspecialized" an eye should be—or can be. How does Mandelbrot's testimony relate to his use of pictures and theoretical thinking, and what clues are provided by his working materials? At this point, it is necessary to return once again to the profound clues provided by his early reference to the Island of Doctor Mandelbrot, and his interest in islands more generally. Pictures of islands and coastlines were among the most numerous finds in his office in 2011. This was not a coincidence. They were crucial to his ability to bridge the gap between mathematical figures and natural forms that had been created even before it became possible to generate images with a computer. The properties of the edges of islands had already been featured in an article he published in *Science* in 1967 entitled "How Long Is the Coast of Britain?," which is now considered the founding text in fractal geometry.[61] There, referring to the "monster curves" of the nineteenth century, Mandelbrot argues that coastlines are self-similar and infinite in mathematical terms, constantly revealing jagged new details under enlargement. In the article, he continues to refer to such structures as "fractional," a term he had adopted from the eclectic mathematician Lewis Fry Richardson, whose findings he discusses in the article.[62] It would not be until the early 1970s that Mandelbrot would have access to the technical machinery he needed to generate pictures from these formulas. Looking back, Mandelbrot often described how the sight of these first computer-generated shapes was first perceived with a shock of recognition that operated in a similar manner as a déjà vu:

> It is impossible to forget the sleepless nights we spent producing the pictures with programs that we had diverted from their intended purpose so as to be able to use them at all. . . . Although we hadn't expected it, relief set in: the curve described the coast of New Zealand and Bounty Island.
>
> [T]he device was very cumbersome, but when the shapes came out, what a revelation! The pictures were so poor. . . . Even so, there was an overwhelming feeling that what we had drawn was right. [63]

Reports such as this document the "Columbus feeling" frequently referred to in descriptions of early computer use.[64] In the years of the first experiments with fractal coastlines, producing pictures took all night, so that Mandelbrot's "speckled geographies" were often thrown away the next morning by computer-room operators who thought them to be misprints. Moreover, for a long time, it was only possible to generate lines, the edges of which were composed of the letters O, X, M, and W; uniform surfaces were not possible. Indeed, some of the islands in the catalogue were later amplified by hand (see Found Images I, cats. 1–4). Mandelbrot did ultimately publish the black, irregular specks in figure 16 in his first book, *Les objets fractals* (1975). He distinguished between them on the basis of their fractal dimension values 1.1, 1.3, and

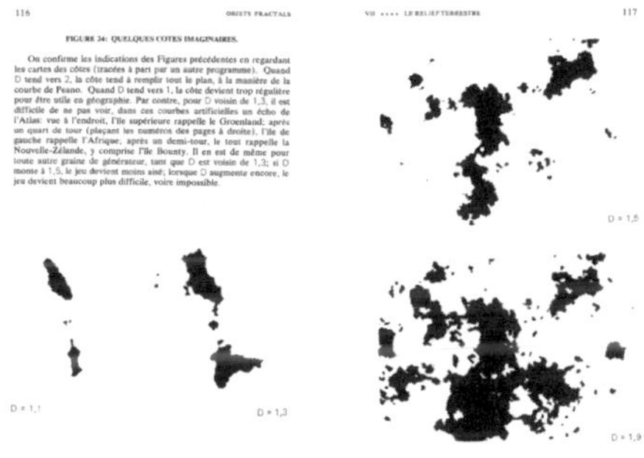
116 OBJETS FRACTALS

FIGURE 24: QUELQUES COTES IMAGINAIRES.

On confirme les indications des Figures précédentes en regardant les cartes des côtes (tracées à part par un autre programme). Quand D tend vers 2, la côte tend à remplir tout le plan, à la manière de la courbe de Peano. Quand D tend vers 1, la côte devient trop régulière pour être utile en géographie. Par contre, pour D voisin de 1,3, il est difficile de ne pas voir, dans ces courbes artificielles un écho de l'Atlas: vue à l'endroit, l'île supérieure rappelle le Groenland; après un quart de tour (plaçant les numéros des pages à droite), l'île de gauche rappelle l'Afrique; après un demi-tour, le tout rappelle la Nouvelle-Zélande, y comprise l'île Bounty. Il en est de même pour toute autre graine de générateur, tant que D est voisin de 1,3; si D monte à 1,5, le jeu devient moins aisé; lorsque D augmente encore, le jeu devient beaucoup plus difficile, voire impossible.

Fig. 16. Benoît Mandelbrot. "Quelques côtes imaginaires." From *Les objets fractals: forme, hasard et dimension* (Paris: Flammarion, 1975), 116, 117.

1.9. Of the shapes identified with D=1.3, he commented that they were the ones that displayed the greatest simulative power. Their similarities to cartography were hard to overlook: The top island was reminiscent of Greenland; with a quarter turn, the one on the left looked like Africa; and after a turn of 180 degrees, one could see New Zealand with Bounty Island in the mass of specks.[65] In that same year, Mandelbrot published similar chance shapes in another context, noting that he had now discovered Greece, the mirrored Sea of Okhotsk, the Gulf of Siam, and western Scotland.[66] Mandelbrot relied on an associative interpretive logic in dealing with such ambiguity, cheerfully rotating his images in search of familiar shapes.

Mandelbrot's interpretation of chance artifacts can be related to a pre-digital history of seeing in the realm of art. His method was preceded by a long tradition among artists of exploiting the creative and inspirational potential of chance patterns.[67] Prominent examples of this practice can be found as early as the Renaissance: In describing the proper practice of artists, Giorgio Vasari refers to the early stage of composition when artists would make a "first drawing . . . in the form of a blotch," which should be considered only as "a rough draft of the whole."[68] He advises that such blotchy sketches (Italian: *schizzi*) be "hastily thrown off" as a means of testing "the spirit of that which occurs to him."[69]

A pictorial technique comparable to Mandelbrot's associative interpretation of his visualized mathematics can be seen in eighteenth-century landscape drawings. The English artist Alexander Cozens (1717–86) first introduced his method of "blot drawing" in 1759.[70] It allowed one to develop landscapes out of arbitrarily placed and presumably nonobjective specks of ink blots (fig. 17). The "blot" itself was "no drawing, but a collection of accidental shapes out of which it is possible to make a drawing."[71]

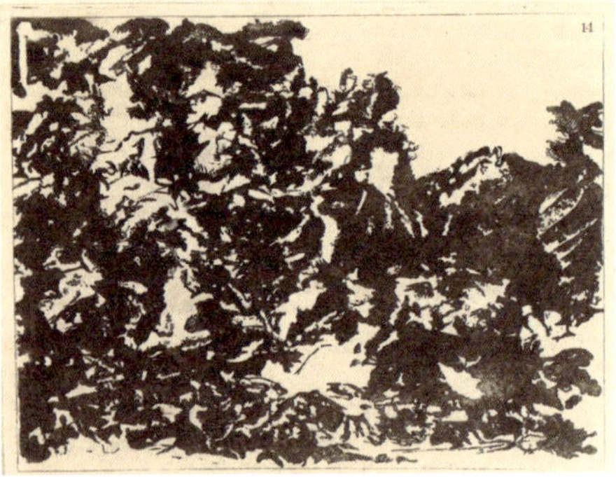

Fig. 17. Alexander Cozens. *A new method of assisting the invention in drawing original compositions of landscape*. Plate 14, described in the accompanying table as "'a blot of the fourteenth kind of composition of landscape'; reproduction of an ink blot made to invent a landscape, suggestive of receding hills or other formations on either side of a plain, 'the horizon above the bottom of the view.'" 1785–86. Aquatint, printed by J. Dixwell, published by James Dodsley. © The Trustees of the British Museum.

Comparing Cozens and Mandelbrot, one first notes their differences: Mandelbrot's silhouette-like, large-scale specks of color come to resemble monolithic, menacing boulders when compared to Cozens's dynamic brushstrokes, which reflect the play of light and shadow and evoke perspective. Whereas Mandelbrot studied his specks from a bird's-eye perspective, Cozens gazes into a three-dimensional landscape—even when the illusion of space is largely absent in this self-contained scene with "receding hills" and a sparse sky, as the title suggests. And yet there are astonishing similarities between the two. Cozens's "blots" were thought of as "multifunctional" elements: "One and the same 'blot' can lead to different sketches" and consequently serve as the starting point for different imaginary landscapes.[72] Landscapes developed in this way were to be formed "out of the spirit of nature and according to the principles of nature instead of merely replicating a portion of nature already formed."[73] Cozens's "natural" landscapes could not be identified with any specific locality, and in this respect his associative picture logic, based as it was on the interpretation of chance blots, was comparable to Mandelbrot's practice. While Cozens aimed for a "reassessment of the relationship between art and nature,"[74] Mandelbrot was suggesting new relationships between nature and mathematics.

That Mandelbrot's images only vaguely resemble the silhouettes of certain countries was part of their message. They were not intended to emulate any specific geography, but rather to evoke the idea or basic principle of nature—to be "true to" it. Indeed, dissimilarities and discrepancies between picture and object were necessary to set him thinking and to construct the "natural fractal." This corresponds to a basic quality of epistemic images in general as analyzed by the historian of science Hans-Jörg Rheinberger, who has stated that both the "ambiguity of the representational process" and a certain "vagueness" of the object were required of such images.[75] By means of his associative method, Mandelbrot was able to access a realm of free thought. Only through the image and its interpretation were completely *new meanings* generated. It was these that produced the connection between geometry and nature, which would not have been made at all without the image as intermediary.

But it was not computer graphics in the first place that suggested to Mandelbrot that he look at chance distributions and nature together. A good ten years before he had access to such technical facilities at IBM, he had already begun to train his eye by looking at traditional hand-drawn statistical representations. With regard to figure 18, for example, he noted in 1963 that if one ignored the picture's source it was possible to imagine that one was looking at a "new world."[76] The image in question, an "experi-

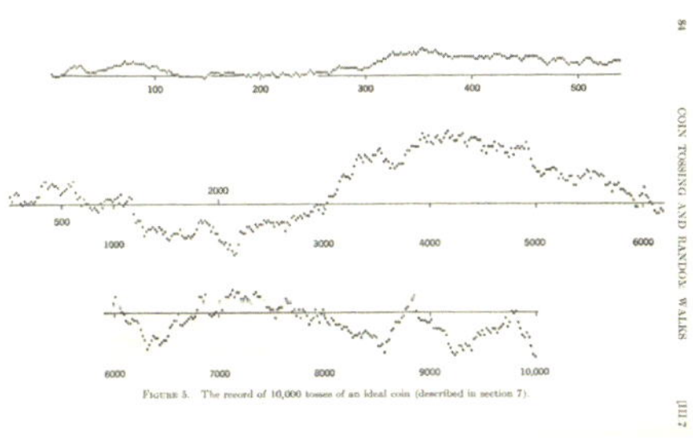

Fig. 18. William Feller. "The Record of 10,000 Tosses of an Ideal Coin." From *An Introduction to Probability Theory and Its Applications* (New York: John Wiley & Sons, Inc., 1957), 84. Reproduced with permission of John Wiley & Sons, Inc.

mental illustration" that plotted on three scales the results of a sequence of coin flips, came from the two-volume textbook *An Introduction to Probability Theory and Its Applications* by the mathematician William Feller.[77] Mandelbrot borrowed Feller's illustration without a single change, though he proposed a geomorphological description of it that added an additional level of meaning: "First of all, forgetting the origin of that figure, let us imagine that it is a geographical cross section of a new part of the world, in which all the regions below the bold horizontal lines are under water."[78] By using another mathematician's picture for his own purposes, Mandelbrot was able to establish a strong and characteristic connection between chance and nature. The chart came to assume an iconic status for him, and he reproduced it in all three of his books on fractal geometry, as well as in a number of articles.[79]

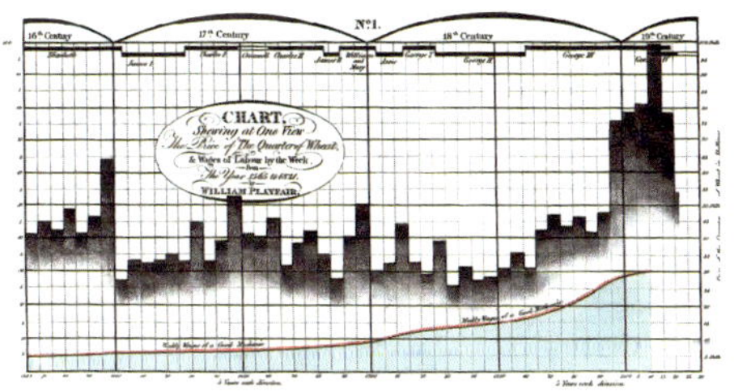

Fig. 19. William Playfair. "Chart Showing at One View the Price of the Quarter of Wheat," 1822. From Edward R. Tufte, *The Visual Display of Quantitative Information* (1983; Chesire, CT: Graphics Press, 2001), 34.

That Mandelbrot's way of seeing served to update visual traditions can be discerned in the development of statistical graphics. For example, the now legendary diagrams of the economist William Playfair, considered one of the inventors of the field, were designed to be easily scanned, a quality characteristic of landscape depiction as well.[80] His Wheat/Wages Diagram from 1822 (fig. 19), for instance, relates the price of grain to the weekly wages of workers between 1565 and 1821. That Playfair wanted the viewer to register the information with ease is indicated by the last three words of the phrase he confidently placed in the middle of the image: "Chart Showing at One View." The synoptic qualities of the visualization were crucial to him, as can be seen in his choice of a background for the changes in wages: the trope of a silhouetted towered city, familiar from early modern cartography, yet colored in a kind of gray sfumato to reinforce the distance. To obtain the impression of a landscape, he combined graph and column diagram, two modes of representation that are very different. Playfair unambiguously if tacitly answered the importance of the "specialization of the eye": In learning to read new picture forms, it is necessary that one be presented with well-established, familiar visual impressions. Instant legibility—showing at "one view"—all but requires the use of other genres' pictorial logic. It is in them that preexisting visual habits are exercised and a picture's subject can be accepted at first glance as logical because familiar. A similar representational mode was employed in the development of geology as a separate scientific discipline at around this time. Indeed, a glance at the chart may suggest geologists' characteristic transformations of mountainous regions into panoramic cross-sections[81]—just one among many examples in history of a mode of visualization being handed along from one science to the other.

Playfair's strategy shows that the eye can accommodate itself to a new picture form if it comes into contact with familiar ways of perceiving reality, as in an image that has been designed in a particular way or, as Mandelbrot demonstrated in his reading of Feller's diagram, an existing image that has been cleverly re-interpreted. Mandelbrot's repurposing of Feller's diagram allowed him to visualize chance in the jaggedness of the earth's surface; in turn, this allowed him to treat the features of a certain part of nature—the "sculpture of mountains"—as something that could be analyzed using statistics.[82]

Both Mandelbrot's first impressions and the later configuration of his images combined meanings and visual traditions from different contexts. These not only influenced the dissemination and popularization of Mandelbrot's work, they were part of his exploratory thinking process, shaping thereby his conceptions themselves.

As for the use of images in the construction of a fractal "geometry of nature," Mandelbrot's "unspecialized eye" sought to figuratively reinterpret representations of abstract knowledge. Schemata that could no longer be deciphered on the experiential level were approximated to visual habits; thus they functioned as an experimental field for new constructions. Indeed, such readings could take on a significance of their own that affected Mandelbrot's interpretation of their content.[83] Mathematical structures that were unrecognizable in visual terms, and accordingly appeared to exclude the eye as an instrument of understanding, were objectified, in analogy to geomorphological features, as visually identifiable spaces.

The way Mandelbrot saw and interpreted his images tended to break down traditional distinctions between the use of pictures in art and in science whereby images in science are supposed to be unambiguous.[84] Instead, as Mandelbrot helped to demonstrate, pictures in the sciences have to be seen as fundamentally obstinate, intractable instruments whose vagueness can be productive in research. Scholars such as Mandelbrot have utilized these qualities of the visual in a playful, associative manner, one that has led them to new insights.

Many of the first artists to experiment with the computer in the 1960s sought to use scientific methods, "with their claim to objectivity, transparency, and verifiability," to reconstruct "an aesthetic on a scientific basis."[85] Mandelbrot took the opposite approach; he made use of artistic techniques that exploited the subjectivity of visual impressions to create natural forms out of images of chance. In this way, a kind of "role reversal" occurred between motives conventionally attributed to the realms of art and science respectively.

At the same moment, this apparent breakdown between traditional distinctions also led to new confusions. One has to wonder about the degree to which this perceived "role reversal" influenced the scientists' self-image. Here it is worth noting a seemingly insignificant detail in some of the pictures in the exhibition: Mandelbrot's signature. On my first visit to his office in 2008, he insisted on signing all the pictures, individually, and with a fountain pen, before he allowed me to scan them for my research—seventy in all. This procedure offered clear evidence of a pronounced sense of authorship—even for those pictures that had been misprinted, torn, or stained (see Found Images I, cats. 41and 43).

Signing a picture is a practice mainly associated with fine art, a way of claiming responsibility for one's *oeuvre*.[86] Mandelbrot candidly admitted that he played with this ambivalence: "For *Who's Who* I put down MATHEMATICIAN AND SCIENTIST or even MATHEMATICIAN, SCIENTIST, SCHOLAR, AND ARTIST, which upsets editors, who prefer a single word."[87] How did he come to think that? Was this mere hubris or evidence of a misunderstanding?

There would be no point in suddenly declaring Mandelbrot's signed pictures to be "artworks." The question itself is unproductive, as it tends to lump together the varied pictorial products of art and science and seduces one into "blurred simplifications."[88] More to the point is whether a separation between mathematics and art is justified at all with regard to pictures and their possibilities—particularly their epistemic possibilities.

Mandelbrot's islands provide a perfect example of how his creative method left its mark on the world of mathematics. Having spent years testing the effects of various dimensional values in order to determine how he might achieve the greatest simulation of "nature" (see fig. 16 and Found Images I, cats. 1–8), Mandelbrot arrived at an important mathematical thought solely on the basis of a visual impression. Looking at some of his pictures, preparing the manuscript for what was to become *Fractal Geometry*, he once again had the sense that he was looking at landforms (fig. 20) and then all of a sudden, "I screamed: Stop it, stop it! . . . I see an island, I see a continent. It's amazing—this is Spain, this is Great Britain . . . and then Scandinavia!"[89] Since he had previously used the fractal dimension of 1.3 for such distinct impressions of nature, he now conjectured that these pictures must also have the value 4/3 (rounded to 1.3). That Mandelbrot's intuition was correct was ultimately proved in 2000 by the mathematicians Gregory F. Lawler, Oded Schramm, and Wendelin Werner.[90] This gap between hypothesis and proof was not untypical of Mandelbrot, whose creative, associative way of looking led him to conjectures rather than to rigorous mathematical statements. Lawler himself compared Mandelbrot's approach to that of the archaeologist who digs and makes exciting discoveries but leaves it to others to provide the explanation or the context—or in Lawler's words: "If he had not produced the questions, other people would have not found the answer."[91]

The creation of visual analogies—the pursuit of analogical seeing—was the engine that drove Mandelbrot's research. Indeed, he came up with the most appropriate symbol for this method himself. In 1974, having set about creating his own designs for the cover of *Les objets fractals*, he began to cut out illustrations from the book and paste them as a collage on its front cover (fig. 21). He produced three separate versions of the design, which was not in fact used, based on his model of the distribution of galaxies, his speck-like "islands," and his "imaginary continents." True to Mandelbrot's credo—"before beginning to

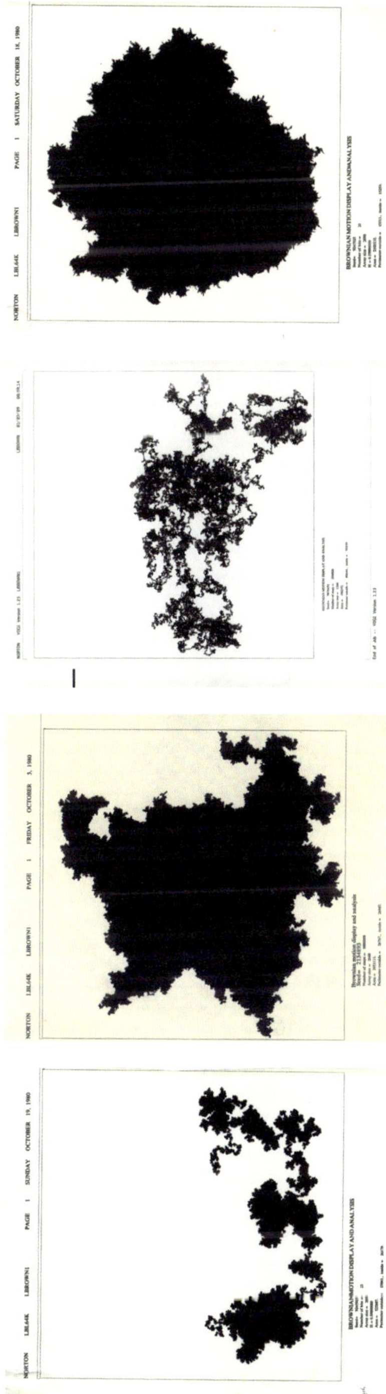

Fig. 20. Benoît Mandelbrot. Brownian motion, 1980–81. Computer-generated prints. Collection Aliette Mandelbrot.

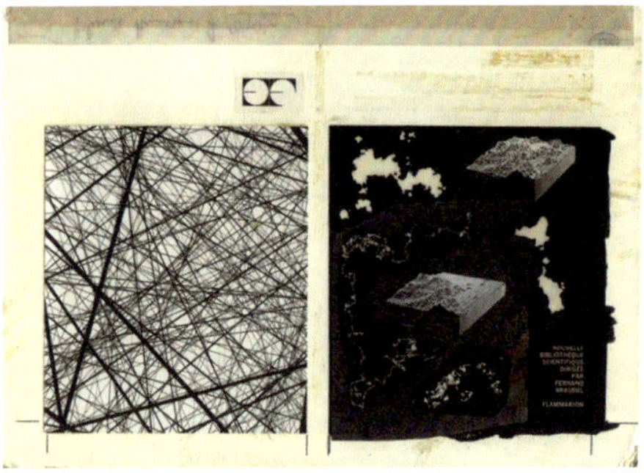

Fig. 21. Benoît Mandelbrot. Cover design for *Les objets fractals: forme, hasard et dimension*, ca. 1974–75. Collage of paper cutouts on cardboard. Collection Aliette Mandelbrot.

understand what fractals are, one should know what they look like"—an overview of the phenomena was to precede even the opening of the book.[92] On the back cover, a collection of straight lines—a basic element of Euclidean geometry—was to appear. Mandelbrot identifies this collection of lines with Parisian streets[93]—but if it is an imaginary city map, it looks as if it had been struck by an earthquake. The viewer feels as if he or she is looking through the lines, which are of various widths, into a box of great depth, while the direction of the streets has been whirled into a confusion that extends across the entire sheet, an effect generated by the application of calculated pseudo-chance. Thus, a stochastically transformed "culture" and a programmed "nature" are juxtaposed and interrelated. In the jacket text, Mandelbrot states: "[I]t will be suggested that there are far-reaching analogies between all these objects that are intuitively comprehensible. Despite their novelty, this must be seen as fundamental."[94] Mandelbrot himself provided the key word there: "analogy." This was a concept fundamental to his thinking. The juxtaposition of images on his book-cover designs recontextualized his picture shapes and gave them new meaning. More generally, the tableau as a configuration of shapes in a common picture space serves to symbolize his entire approach. This gives a first answer to the initial question of how mathematical thinking can be organized and make sense given a surfeit of similar pictorial material in a series. The single picture showing (or rather *embodying*) the exemplary individual case becomes more general—demonstrating a way of seeing that creates analogies and reveals associative potential as part of a multitude—while still keeping its validity as a unique image. In short, a logic of seeing takes hold that can be described as the "logic of the tableau."[95] But it should also be noted that with his theory that computer graphics alone had led to a "return of the eye" in mathematics, Mandelbrot underplayed the role of sense impressions in the acquisition of knowledge and in so doing helped to skew people's assessment of his own contribution. On the contrary, it was his imagination, initially developed independently from the computer, that had inspired his research and served as an antidote to the dominance of calculation.

The Emptiness Between Islands: Media and Visibility; or, The Computer and Its Blind Spot

I saw it but I didn't know that I was seeing it.[96]
—Benoît Mandelbrot

How visible are islands? Does an island stop where the water begins? And how can one discover what lies between islands? Such questions may seem strange at first, yet they lead to the crux of a riddle that was still unsolved in 1980—and to a second way that islands were important to Mandelbrot's pictorial thinking. From imagination and the figurative meaning of the islands as geomorphological structures, the focus is thus shifted to their abstract counterparts: "islands" and "continents" as metaphorical terms that Mandelbrot used for decisive pictorial details on his computer-generated prints when he was about to make one of his most well-known discoveries, the Mandelbrot set. These components, and especially the mysterious spaces between them, raised serious issues about the mutual dependency of the world of mathematics and the image technologies that generated them. It was the riddle of those islands and the question of whether they were really peninsulas that was plaguing Mandelbrot and that revealed the twofold, antagonistic power of the visual for thought: the potential to seduce the mind through mere appearance. As the pictorial history of this famous symbol of chaos theory demonstrates, vision and thought had by no means always been perfectly in harmony. Rather, seduction can be determined as a second force that coexists with imagination. Imagination and seduction are two sides of the same coin in visual thinking.

The majority of the pictures from Mandelbrot's office on display in the exhibition are related to experiments he carried out in 1979–80, shortly before his discovery of the set (see Found Images I, cats. 9–39). Yet if these images are dominated by mere blankness, patterned by a covering of small dots, or structured by fuzzy shadows or indefinite outlines, the Mandelbrot set became generally known in a wholly different form, dazzling and brightly colored, as in the computer graphic that appeared on the cover of *Scientific American* in August 1985 (fig. 22), which looked like a dynamic fireball racing down a black slope. Above all in the margins, the Mandelbrot set is distinguished by impressive ornamentality and self-similarity. This is all the more striking given that the formula it is based on seems simple even to non-mathematicians.[97] This picture is typical of the popular composition and distribution of the set, which in the mid-1980s became one of the first scientific "icons" in the digitalization era.

Despite its shaded, vivid, and relief-like appearance, the Mandelbrot set is at heart nothing but a series of numbers that can be graphically represented within the complex number field. Because complex numbers consist of two components (real and imaginary), the picture space of a digital graphic that sits in a coordinate system with its x and y axis allows us to produce a correlation between the invisible equation and the visible surface: Each pixel on the screen corresponds to one precisely defined complex number. That said, the discrete, grid-like structure of digital-visualization

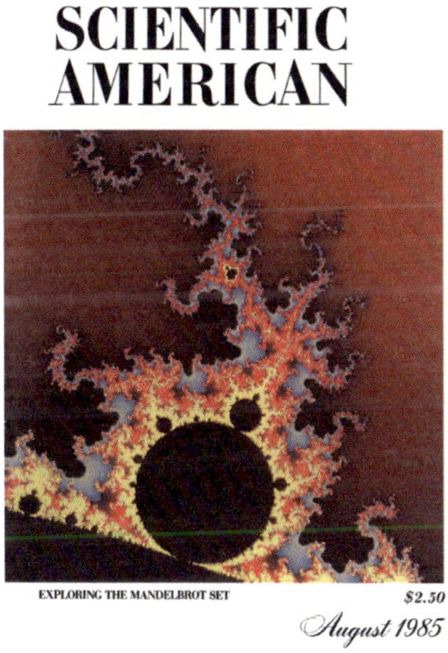

Fig. 22. *Scientific American* cover, August 1985. Reproduced with permission. Copyright © 2012 Scientific American, Inc. All rights reserved.

technology has also generated its own zones of insecurity for this correlation: While a nonlinear dynamic structure generally takes up an infinite and continuous space, if it is to be represented digitally, on a computer, it has to be reduced and approximated. As a consequence, pictorial details that seem unrelated or "far apart" on an image may well be connected or "very close" as far as their provable mathematical properties are concerned. In short, visibility and existence in a mathematical sense—what can be seen and what can be rigorously proven—are not the same.

The gap between appearance and "reality" became a serious challenge for Mandelbrot's investigation of his "mathematical islands." On the aforementioned 1985 cover (see fig. 22), for example, there appears a black bulbous shape that Mandelbrot would have called a "continent" (i.e., the main, central part of his set); on the upper level, in a black squiggly blaze, one can make out a black shape of similar appearance, only smaller: an "island" (i.e., a self-similar replica). In this image, there are no empty spaces apparent between islands and the continent. Both are connected visually through various color nuances and shades. But this was by no means the case during the discovery process. At first, continents and islands were separated by empty spots. What was the significance of the transition from empty intervals to color connections? To explain the mystery of the islands and their problematic visibility—and hence their image-theoretical relevance—it is necessary to take a closer look at the process by which they were discovered.[98]

The discovery of the Mandelbrot set was based on work done by the French mathematicians Gaston Maurice Julia and Pierre Fatou at the beginning of the twentieth century on the dynamics of polynomials in the field of complex numbers.[99] Their papers were held to be extremely difficult masterpieces and as such were long considered all but impossible to understand, even by mathematicians. Mandelbrot studied under Julia at the École polytechnique and knew his work (see fig. 4, page 158), but until the late 1970s, he had never concerned himself closely with it.[100] It was only when he realized that a computer might be used to produce pictures from Julia's analyses that his interest was awakened. At a colloquium in Princeton in 1977, Mandelbrot was made aware of computer graphics by the mathematician John Hubbard, with whom he made contact a year later.[101] Hubbard was one of the very first people to produce graphic depictions of cubic polynomials (fig. 23). Though Mandelbrot spoke of Hubbard's pictures as being "of rather poor quality,"[102] they must have made a strong impression on him, as he soon invited Hubbard to IBM to learn about his ability to generate computer visualizations from Julia's formulas. This was not long before he himself, armed with some of the most powerful computers of the time, began producing pictures of so-called Julia sets, a practice that would ultimately lead to the definition of the Mandelbrot set.

It is possible to distinguish two phases in Mandelbrot's research. In the first, at IBM in 1979, he devoted himself exclusively to the graphic depiction of complex non-quadratic equations. In the second, as a visiting professor at Harvard during the winter semester of 1979–80, he focused

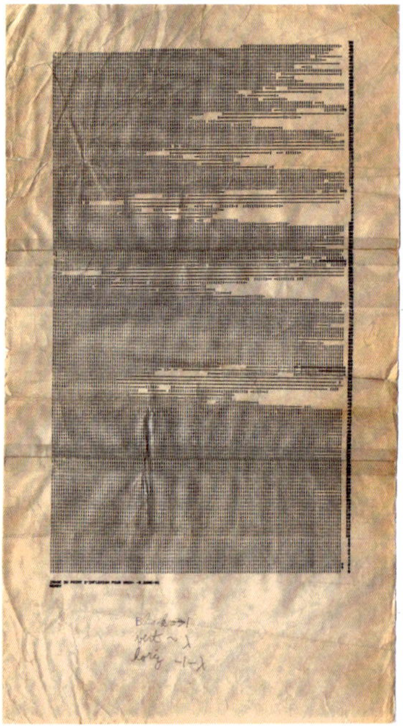

Fig. 23. John Hubbard. Investigations in the complex plane (Newton's method), ca. 1977. Computer-generated print. Courtesy of John Hubbard.

under less than ideal technical conditions on quadratic equations and their illustration. The two phases can be characterized in terms of sense perception, the first being a *phase of seeing* and the second being a *phase of discovery*: "I did see those elements [i.e., what were later called the small "islands" as part of the Mandelbrot set] in 1979 but could not organize and describe them. Therefore, they remained undiscovered."[103] That his seeing did not take place at the same time as his discovery contradicts the all-too-simple assumption that the Mandelbrot set simply suggested itself spontaneously once calculating capacities reached a certain level. Such a view is no more accurate than the claim that the set's discovery would have been possible before the invention of the computer. The history of the Mandelbrot set can neither be disassociated from the requisite technology nor attributed to it alone. The main difficulties that stood in the way of its discovery at the beginning resulted from Mandelbrot's having to deal with pictures new to this field, pictures that needed to be observed. With the computer, observation became a new epistemic category of mathematics.

During the phase of merely seeing and of playful experimenting with the new shapes, Mandelbrot, together with his programmer at the time, Mark R. Laff, produced an extensive series of black-and-white graphics. During my survey of Mandelbrot's office, I came across some 150 images from this period which he characterized first as "state-of-the-art," then as "used furniture," and finally as "valuable antiques."[104] They document most impressively how comprehensive his experiments were and how difficult it was to make anything visible at all. Mandelbrot did not publish these pictures at the time because it was simply not possible to make sense of them or to "clean up their theory."[105] At the beginning, it was by no means certain that any of his experiments would create a clearly recognizable pattern at all. An equation's complexity was often reflected in an apparitional blurring of its contours—or even, as happened frequently, in the fact that the picture simply remained white, and nothing at all could be seen. Some of the images are immediately reminiscent of glimpses into distant galaxies (see Found Images I, cats. 18–21). The appearance and disappearance of the patterns seem to be synthesized in a single picture space.

Whereas the choice of coloring is arbitrary and depends on the technology used, the assignment of colors—or the decision to just use black and white—has to follow neatly defined rules that have to be fixed before the image is generated. These rules identify the range of values—the threshold—for which a specific color will be assigned. To generate a graphic, the computer takes each of the numbers in the plane (i.e., pixels), inserts its value into the formula, and iterates it for a previously defined number of times to find out what specific color has to be assigned. In chaotic dynamics, it is a characteristic of the equations to react strongly even to the tiniest changes, a quality known as "sensitive dependence on initial conditions." Consequently, small changes in the parameters of an equation can lead to completely different behaviors. The technological limits of the resolution entail two major problems. First, two patterns that are theoretically distinguishable may blur in the computer graphics and thus appear to be indistinguishable. Secondly, errors in rounding-off induce artificial changes in state, which in turn influence behavior in the long run because of the equation's dependence on initial conditions. Previously distinct shapes and patterns may, therefore, start to fade and blur. The technology and its limits of resolution determine the degree of fuzziness.

The difficulty of achieving precision in the representation of equations explains why, in many of the images—especially the huge series of 150—only a few outlines are clearly recognizable, with the majority disintegrating into irregular cloudlike spots: "Throughout," wrote Mandelbrot, "I felt like someone trying to photograph the Cheshire cat in *Alice in Wonderland* at the very moment it is about to disappear."[106] The pictures that he compared to the transitory appearance of the famous mythical feline from Lewis Carroll's novel reveal the fragile condition of the number field known among mathematicians since 1918 as "Lattès chaos" (see Found Images I, cat. 35).[107] The closer the parameter value λ approached the number 1, the larger the gray areas in the picture became.[108] Like the photographer who controls his shot by his choice of lens and his use of lighting, Mandelbrot identified more or less photogenic and picture-worthy motifs in the complex number field. Instabilities between appearance and non-appearance are indicative of the fundamental characteristic of non-linear equations. Because of their great sensitivity to changes in parameters, they lend themselves to the illustration of the transition between order and chaos in dynamical systems. What was to be considered order and what chaos depended on the visual arrangement: For example, gray areas with interference, from which a pattern was wrested out of multiple variations of the most diverse parameters, belonged to the classic iconography of chaos. Mandelbrot acknowledged in retrospect that in places in the series it might already have been possible to recognize the familiar bulbous shape of the Mandelbrot set. He subsequently glorified these as "advance-shadows of the Mandelbrot-set-to-be,"[109] which he had not yet identified as a shape of its own. (See similar structures in Found Images I, cats. 18–21).

The blurring and the empty spaces in these pictures, however, point to the complex relationship between formula and image. Visualizations can always only be approximate values of non-linear sets. Since the calculations inevitably have to be performed with a finite number of iterations, the resulting images always bear traces of a compromise. The encounter between formally infinite mathematics and the finite abilities of the computer owing to its limited processing power necessarily produced indefinite and hence blurry areas. As a result, Mandelbrot focused on investigating appearances and disappearances and searching for the right picture moment for recognizing a pattern, as well as on the different possible meanings of the empty spaces on the sheet of paper. It was not always possible to determine unambiguously how the visualization and the mathematics related to each other. Owing to the technique and its limitations, there were built-in contradictions and areas of friction between theory and image. It was because of those interstices and empty spaces that the island issue would soon become acute—and tempt Mandelbrot into hastily drawing mathematically unproven conclusions.

The problems of visualization became especially critical when Mandelbrot moved to Harvard. There he had neither an office of his own nor his own machines to work with. He was dependent on the older computers in the basement of the Science Center that were available to everyone, shimmering green monitors with cathode-ray tubes and dated printers. The brief working time available and comparatively poor equipment forced him to reduce the complexity of his research subject. Turning away from cubic polynomials, he decided, together with his assistant Peter Moldave, to experiment with quadratic polynomials with a single variable parameter. What he expected was a simple, unambiguous result. Yet what appeared was unlike anything he had experienced before. The first pictures "looked awful: filled with apparent specks of dust that the Versatec printer produced in abundance. . . . I was motivated to sniff out the ramifications of those specks of dirt" (fig. 24).[110] According to all reports, the monitor at Harvard was subject to constant malfunctions and the decrepit printer made it uncertain what one was really seeing. Were the specks in fact dirt or "dust"? Were they smears and blots of the sort that the machine produced in abundance or visualizations of mathematics? For the moment, the boundary between image and accident was blurred. Suddenly, something surfaced that could not be understood and that raised new kinds of questions: Were the specks part of the picture or not?[111]

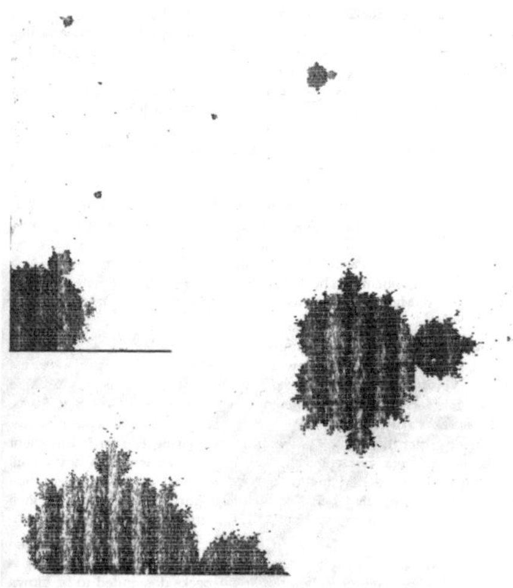

Fig. 24. Benoît Mandelbrot. "My first picture of the whole M set is—unfortunately—either misfiled or lost. [...] The top panel here is a blow-up of the most conspicuously 'messy' corner of M, near the bifurcation of order 3. The middle and bottom panels show the oldest preserved blow-ups of the two largest islands, one already seen in the top panel and the other intersected by the real axis (for reasons of economy, only half was computed)." Harvard, Spring 1980. From Benoît Mandelbrot, *Fractals and Chaos: The Mandelbrot Set and Beyond* (New York: Springer, 2004), 13.

In order to distinguish between picture and phantom images, Mandelbrot decided to examine the specks more closely, employing in effect the computer as a microscope. It was only his perception of symmetries in the "dirt" that elevated the "reality probability" that the specks belonged to mathematics: "Since bona fide dirt is not symmetric, both kinds of specks demanded to be blown up for close inspection. . . . The symmetric points had more chances of being real."[112] From a series of enlarged pictures, it was determined that some of the specks were in fact dirt from the machinery, but others proved to be self-similar copies of the basic shape. The presumed glitch was thus transformed into a decisive discovery.[113]

Mandelbrot dubbed the self-similar "messy" speck structure surrounding the main body of the Mandelbrot set as "islands"—"offshore islands"—though sometimes he also referred to them jokingly as "pearls in the pigsty" because they had been discovered in the dirt.[114] For years he tried to ascertain their precise structure. The crucial question was whether or not they were linked to the main structure of the set. The computer images gave them the appearance of free-floating elements. Yet how did the picture relate to the theory? Could their mathematical nature be related to the way they appeared? Were the visible parts of the picture also *existent* in a mathematical sense? Could he believe what he saw?

Because the small specks were deceptively similar to interference or printer smudges, they were repeatedly retouched by conscientious computer-lab and print-shop employees. For that reason, in the years 1980–81, Mandelbrot got in the habit of posting warnings on his output equipment (fig. 25): "Do *not* clean off the dust specks, they are *real*." According to his own reports, the fact that his early fractal pictures were in constant danger of being erased had infuriated him on many occasions. Likewise, his first "chance geographies" (see Found Images I, cats. 1–8) had repeatedly landed in his co-workers' wastebaskets as presumed discards or misprints. With regard to the "reality status" of his visualized

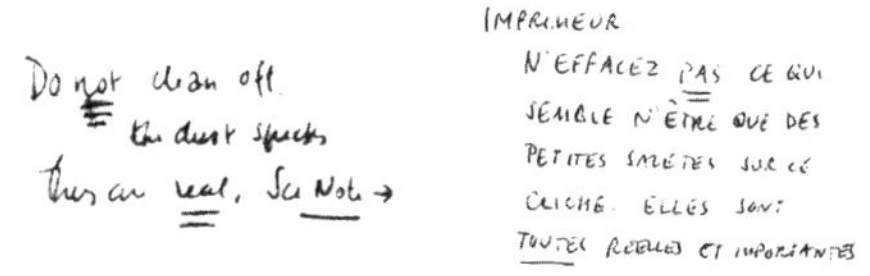

Fig. 25. Benoît Mandelbrot. "Notes on Working Equipment from the Years 1980–81." From Benoît Mandelbrot, *Fractals and Chaos: The Mandelbrot Set and Beyond* (New York: Springer, 2004), 22.

mathematics, a striking reversal can be seen in his attributions: He now called his simulations of natural forms "imaginary,"[115] while the visualizations of his complex equations were classified as "real."[116] Mandelbrot was fond of insisting that he had created his new geometry "out of the dirt"—clearly a further jab at the formalistic Bourbakists, to whom "dirt" was the avowed enemy of mathematical, "crystalline" logic.[117]

Mandelbrot's unconventional and creative reliance on his "unspecialized eye" would never have been possible without pictures. Here, however, it can be seen that his trust in it had its drawbacks: In his first publication of the Mandelbrot set, in 1980, Mandelbrot confused the mere "appearance" of the small islands with their mathematical existence. In his manuscript, he noted: "A striking fact, which I think is new, becomes apparent here: FIGURE 1 is made of several disconnected portions."[118] The specks of islands seemingly detached from the main body of the set were declared, in a fateful confusion of image and proof, as a "fact," although this assertion was clearly unproven, as it was not supported by any mathematical theorem.

Moreover, observant viewers of the picture published in the "Annals of the New York Academy of Sciences" could not even make out the specks discussed in the text, for once again they had been deleted as supposedly unwanted printing flaws by the editor (fig. 26): "Horrors! It is now free of specks! . . . Clearly, gremlins in the printing business had . . . repeated that evil deed."[119] In offprints of the article that he sent to colleagues, Mandelbrot reinserted the deleted picture details by hand (fig. 27).[120] Since he did not see any connecting lines and had

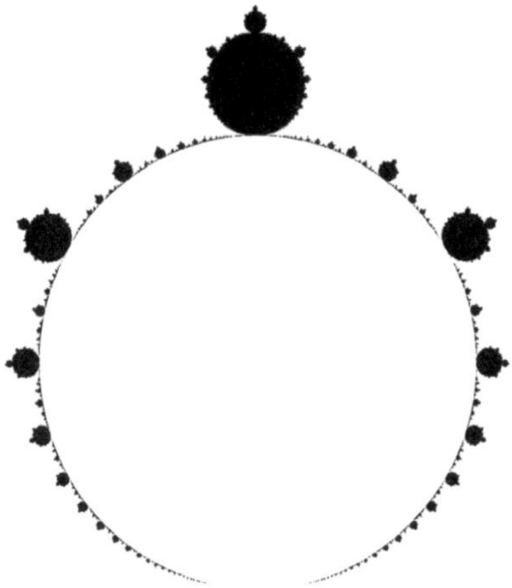

Fig. 26. Benoît Mandelbrot. Complex-plane map of the λ-domain Q. First publication of Mandelbrot set. Surrounding specks (called "islands") had been removed mistakenly during the printing process. From Benoît Mandelbrot, "Fractal Aspects of the Iteration of z → λz (1-z) for Complex λ and z," in *Nonlinear Dynamics: Annals of the New York Academy of Sciences* 357 (New York: New York Academy of Sciences, 1980), 250. Reproduced with permission of John Wiley & Sons, Inc.

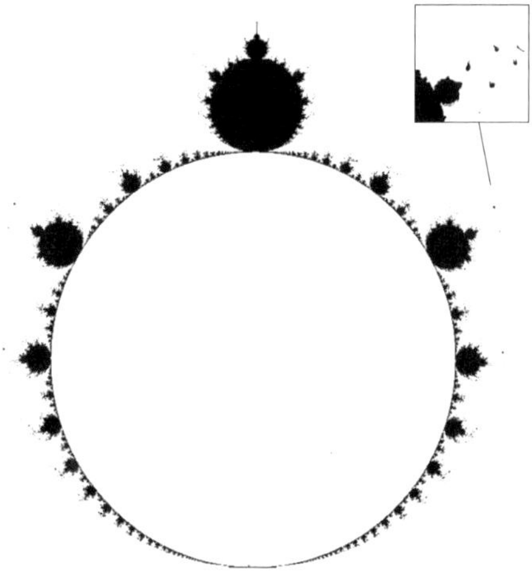

Fig. 27. Benoît Mandelbrot. Offprint of the first publication of Mandelbrot set, 1980, with specks drawn in by hand. Courtesy John Hubbard.

unlimited faith in his computer, he did not believe they existed, a faith he reinforced in the drawing. Suddenly, Mandelbrot's "seeing is believing" had changed to "drawing is believing": to trust in the digitally produced shape, which could be—and had to be—authenticated in drawing to exist. By means of his own successive acts of reinterpretation, the same pictorial details mutated from a hint to a proof and finally ended up as the main result: A material trace became mathematical *evidence.*

The mathematicians Adrien Douady and John Hubbard provided the counterevidence in 1982. In an article in *Comptes Rendus*, they determined that fine lines did connect the specks with the main body but that they could not be made visible at that point in history: "These islands are in fact attached by threads that evade the computer."[121] Douady compared the complex problem of making these structures visible with trying to see tiny mountain rivulets from an airplane flying overhead.[122] When Mandelbrot was making his experiments, it was technically impossible to draw the connecting lines, or filaments, with the existing graphic programs. There was a built-in impenetrability between the two "planes" of the digital image, its visible surface, and the indiscernible columns of numbers that generated the picture (i.e., the operational undersurface that processed the formula).[123] Accordingly, an unbridgeable divide was created between the "islands" and the "continent." A link between the two planes was achieved only a short time later through depictions in shades of gray or colored graphics. Color was able to fill the previously blurry or empty zones at least visually and to distinguish a variety of different behaviors in the dynamical system. In consequence, Mandelbrot felt compelled to expand his earlier slogan, "To see is to believe," into "To see in color may lead to an even higher intensity of belief."[124]

If one were to write a history of the sciences as a history of pictures, one that told the story of the construction of the visible out of the invisible, the issue of limitations on what can be visualized—i.e., visibility thresholds—would necessarily be a primary topic. At what point does something come into view, when does it disappear again, and when does one decide that it is "visible enough" to be perceived as an actual, provable phenomenon—to be judged as a mathematical "fact"? This touches on basic questions concerning the epistemic function of technical images and the elementary relationship between picture-producing methods, pictures, and knowledge. New picture-making methods can open up new realms of visibility and understanding to the same degree that they can also once again limit our seeing.

In Mandelbrot's case, the realm of visibility also defined the horizon of understanding; or, as the science historian Thomas Schlich has written regarding the problematics of microscope pictures, "The limits of understanding coincided with the limits of the visualization technique."[125] Since Mandelbrot placed unlimited trust in his machines and had not seen the connecting lines—could not see them—he had failed to credit their existence and had confirmed as much again in his drawing after the unwanted retouching. His own equation of seeing and believing had misled him into claiming he had discovered a "striking fact" from the mere appearance of the pictures.

Nevertheless, the story of the Mandelbrot set is not one of confusing pictures with reality or of subjective naiveté. On the contrary, it raises the question of how one arrives at understanding through computer graphics in a paradigmatic way. It clearly shows how difficult it can be to maintain a distance between what the eye can see and what can be proved analytically. As the story of a struggle for agreement between observation and theory, it illustrates equally clearly how necessary it is for the eye to be schooled on pictures and how risky thinking inspired by pictures can be. In short, it says something about the fundamental ambiguity of visual perception.

When one considers all of the subsequent events, one comes to recognize that the story of the Mandelbrot set also includes an element of tragedy. Looking back on these events in an article published six years later, Mandelbrot not only regretted the false assumption made in 1980 and the "awkward surrogate"[126] of a conclusion that resulted from it, he also searched for ways to exonerate himself after the fact. One notes this specifically from his retrospective narrative of the story and from the caption beneath his subsequent publication of Julia sets in 2004. There he insists that an analysis of the theory had taught him as early as 1980 that the images of Julia sets were deceptive, and then he adds: "These graphs were important to my thinking because they sufficed to show that the broken-up early M set pictures were compatible with connectedness."[127] It would be a simplification to call this surprising statement a mere contradiction. First and foremost, it emphasizes the intellectual struggles that computer-generated mathematical images were able to cause and the ambivalence that they triggered. Second, it demonstrates an attempt to prove after the fact that his thinking was actually correct despite his previous erroneous assertions. Given the importance Mandelbrot accorded to the virtues of pictures in all of his work, his visually triggered misinterpretation in the first publication of the Mandelbrot set in the "Annals of the New York Academy of Sciences" in 1980 must have sat like a thorn in his flesh.

Dr. Mandelbrot's islands and the issue of their visualization played a key role in his work as a mathematician. On the one hand, they illustrate the virtues of the "unspecialized" eye and thus of the interdisciplinary effectiveness of visual perception. On the other hand, they highlight the seductiveness that adheres to computer-generated visualizations of formulas. In this way, they get to the heart of the question of how pictures are able to convey knowledge and can be seen as symbolic of the two opposing, centrifugal forces involved in pictorial thinking.

The formula "Seeing is believing" fails to give due credit to the particular role that the visual plays in the process of scientific cognition. Pictures have to raise doubts and stir the imagination if they are to generate understanding. In that way they can serve as the raw material for thinking or as thought thresholds for the mind.

The Linear Makes the Non-linear Visible: Drawing Is Seeing

Drawing is a discipline of vision.[128]
—Edward Hill

If one wants to pursue the investigation of the interdependency of the visible sphere and the world of mathematical ideas, it is especially fruitful to continue to look for interactions between early computer-generated pictures and hand drawings. As can be seen in Mandelbrot's work, one of the capacities of drawing is the authentication of computer-generated results. But the potentiality of drawing goes far beyond this function. To identify other epistemic roles of drawing, it is now time to leave the work of Mandelbrot and have a closer look at the practices of some of his contemporaries as a comparison, and first of all, at the work of Adrien Douady.

Not all mathematicians who worked with computers shared Mandelbrot's conviction that computer graphics had "brought back the eye" into mathematics. Douady, a French Bourbakist, an expert in the field of complex dynamics, and the man who first proposed naming the Mandelbrot set after Benoît, was of the opinion that pictures and the eye had never wholly disappeared from mathematics insofar as geometry-oriented mathematicians had been making drawings all along. He reported that even when the dogma of iconoclasm was in ascendance, the visual had found secret yet effective ways into the discipline. Mathematicians had been known to climb rooftops in Paris to draw gigantic diagrams with crayons on walls and to make sub rosa sketches on pieces of paper under tables during lectures.[129] He stressed that "the eye" had only disappeared from the sphere of publications ("expression") but never from the sphere of thinking and the creation of ideas ("conception"). As an experimental tool, the computer had revealed new structures, to be sure, figures that one would never have been able to see otherwise—like all the visualizations of iterated polynomials that Mandelbrot had produced in 1979 (see Found Images I, cats. 18–39)—but that alone had by no means facilitated a *mathematical seeing*: "*Sur cette donnée brute, vous n'y voyez rien*." (On this *donnée brute*, you don't see anything.)[130] The French term "*donnée brute*," which Douady uses for digital pictures both seen on the screen and as printouts (fig. 28), can be translated as "raw data"; it refers to pictorial material that has yet to be processed analytically. A second meaning of the term, the "raw given," makes it clear that here the mathematician is operating from the start not only with data but with something visually concrete.

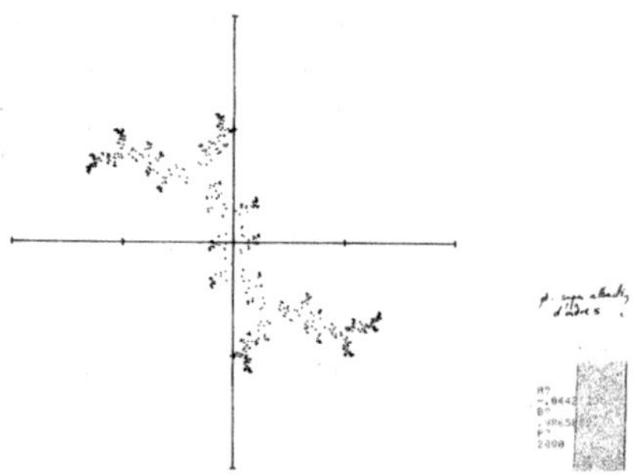

Fig. 28. John Hubbard. Julia set, early 1980s. Computer-generated print. Courtesy John Hubbard.

In their basic building blocks—and because of the low resolution of computer graphics in the early 1980s—representations of Julia sets consisted of exploded black rectangles that only outlined *raw shapes*; they had yet to be made accessible for seeing, at least in the sense of mathematical comprehension. Computer-graphic representations of formulas contained a surfeit of information, which Douady translated in an epistemological three-step process from a digital to an analog medium in order to interpret them. One of Douady's worksheets (fig. 29) shows how he first placed the "raw," "unpolished" rectangles derived from a formula on a sheet for preliminary consideration (on the left), then drew connections between some of those tiny rectangles (in the middle), ignoring others or accenting them with small circles (*interprétation*). In a third step, seen at the far right of the picture, he reconstructed and abstracted

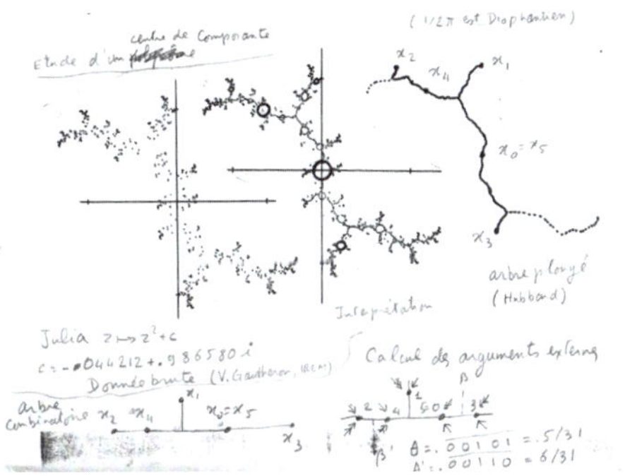

Fig. 29. Adrien Douady. "Hand-Drawn Analysis of a Computer-Generated Julia Set." Manuscript for "Systèmes dynamiques holomorphes," *Séminaire N. Bourbaki* 599 (1982): 51.

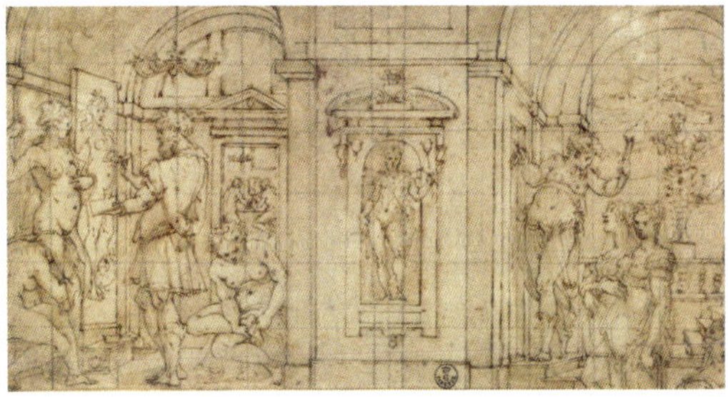

Fig. 30. Giorgio Vasari. Preliminary drawing for *The Studio of the Painter*, after 1561. Gabinetto Disegnie Stample degli Uffizi, Florence, 1180 E.

the information it obtained in a branching line. That line, employed as a tool in the analysis of the dynamics of Julia sets, is reminiscent of a curved branch, sharply bent at the ends (*arbre plongé*), that would later come to be known as a Hubbard tree.[131]

Douady's drawing was an independent tool for understanding that played both a mediating and a constructive-transformative role. The mathematician first explored the computer-generated picture with his pen. Then the independent line itself, no longer embedded in the coordinate system, became a subject for mathematical investigation. In a radical reversal of the Platonic dictum that discursive thinking (διάνοια) requires no visible forms,[132] the drawing became both a tool that helped Douady to understand the computer-supported mathematics of complex dynamics and a mathematical object to be analyzed by his διάνοια.

The three steps from *donnée brute* to drawing recall the function of *disegno* in early modern times, as employed, for example, by the artist, biographer, and theoretician Giorgio Vasari in the mid-sixteenth century in a preliminary drawing for his fresco *The Studio of the Painter* (after 1561) (fig. 30). There the viewer gazes into a pair of barrel-vaulted spaces connected by a corridor and doors. On the right, female models are assembled in front of a multi-breasted Diana Ephesia meant to embody the laws of nature. They are about to step over into the left half of the picture, where a painter is working at his easel, his gaze directed at additional posing models. The center compartment contains a niche holding a male statue, identified by its attributes and triple profile as a personification of *disegno*. Vasari here translated his theory of art, formulated in the second edition of his *Lives of the Artists* (1568), into a concrete pictorial allegory: As a draftsmanly step in the development of an idea, *disegno*, the basis and source of all arts, plays an intermediary role between nature and art.[133]

Douady's conversion of the computer picture into a curving line translated the classic use of *disegno* into a method for generating a mathematical theory. In a "corrective overview,"[134] the mathematician extracts a mathematical concept (*concetto*) from the "raw given" by means of drawing. In both cases, it is only the drawing that makes possible the process of visual transformation. As opposed to Vasari's practice, however, the end product of draftsmanly conceptualizing in mathematics is not a unique work of art or the perfect image but rather a theoretical notion coagulated into a drawn line, a "snapshot" of knowledge, which is fed into the thinking process and used for further deliberations.

Once the telling lines had been extracted from numerical mathematics, they could function as a way of stemming the tide of serial similarities characteristic of images in this field of study. The repetition of nearly identical images can be tiring. Douady relates that he once almost fell asleep in a Paris café in 1982 when presented with

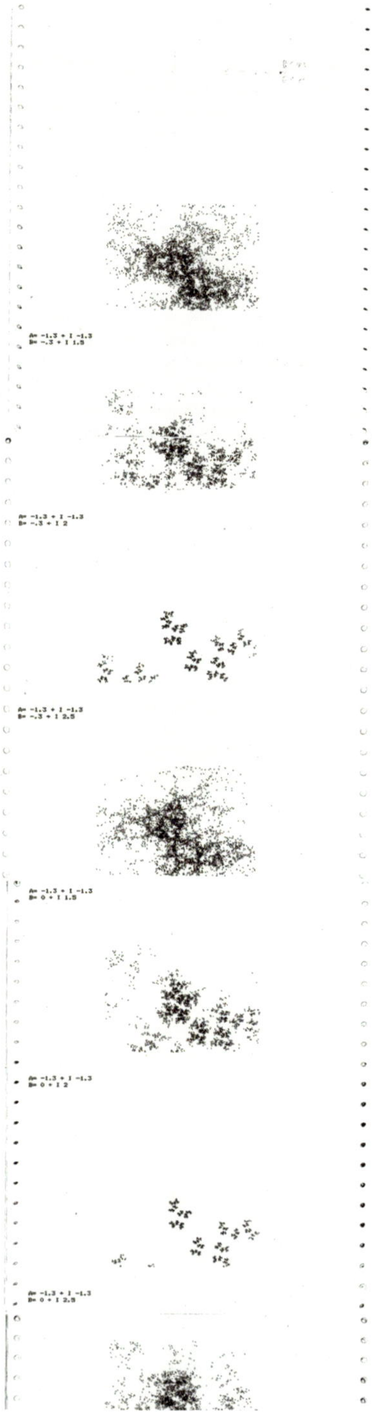

Fig. 31. John Hubbard. Julia sets from rational fractions of the second order, 1982. Computer printouts on continuous paper. Courtesy John Hubbard.

a series of similar pictures composed of "computer-generated clouds of dots"that his mathematical colleague John Hubbard had placed before him (fig. 31).[135] Fresh from the printer on so-called continuous paper perforated at the edges, they represented seemingly endless variations of Julia sets, graphic depictions of rational functions produced by a dot-matrix printer.[136] Hubbard was unable to make out the significance of the pictures and so had confronted his colleague with a whole stack of computer graphics. Douady tells of the encounter in such a colorful way that it is worth quoting him directly: "He asked me: 'Can you recognize anything?' I had nearly fallen asleep, and could only see a succession of clouds that said nothing to me. Suddenly I said to him: 'There, an upside-down rabbit!'—'Where?'—I took an orange marker and traced it in the picture."[137] Douady had observed the succession of similarities, and it was only the recognition of a shape and his drawing that relieved him from boredom. The "rabbit" that Douady felt he could see in the clouds of dots was a playful term for a specific quadratic Julia set, the characteristics of which he had intently studied for years (fig. 32). It owes its name "fractal rabbit" (*lapin fractal*) to what appear to be ears extending from either side of the globular body.[138] In his draftsmanly retracing of the *lapin*, the scientist was treating the digital form as if it were clay or rubber: The rounded center section was taken apart and the "ears" were folded down along the inner edges and placed in perfect alignment with the sides of the printed pattern (fig. 33). Douady had turned the graphics into usable

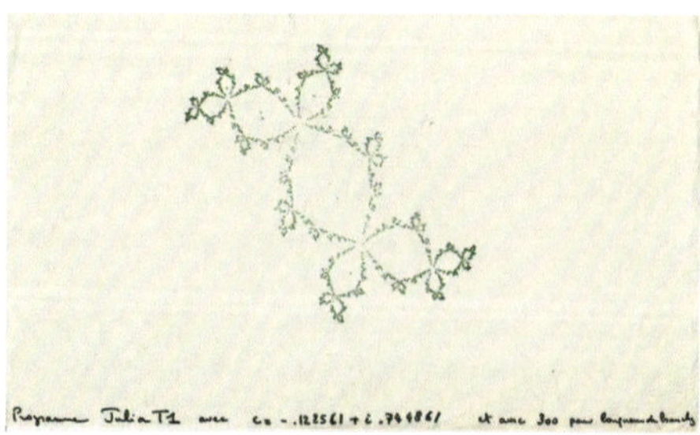

Fig. 32. John Hubbard. Julia set, or *lapin fractal*, early 1980s. Computer-generated print. Courtesy John Hubbard.

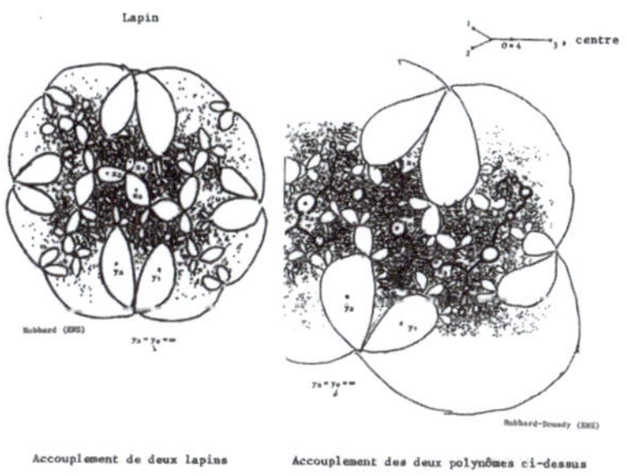

Fig. 33. Adrien Douady. "Theory of Matings (Fractal Pairing), the Encounter of a *lapin fractal* with a Hubbard Tree." From "Systèmes dynamiques holomorphes," *Séminaire N. Bourbaki* 599 (1982): 59.

material for thought. In the process of his trying to understand the data before him, the line derived from the cloud of pixels came to seem like a plastic material, pliable and ductile. While the drawing in the first example helped him to review the basic structure (see fig. 29), he now adapted the branching line of the Hubbard tree as if it were made of wire and could be bent at will (see fig. 33). What was of importance mathematically was the way it converged in the center of the picture with the rabbit's sweeping semicircular arcs and the loops surrounding the image.

Despite its humor and subjective, anecdotal character, Douady's description of his discovery is of central importance in the history of the representation and understanding of complex dynamics. What Douady and Hubbard first discovered together in pictures they were later able to transform into a valid mathematical proof. They gave the name "matings" ("*accouplement fractal*") to their theorem because the mating of two Julia sets of different classes of equations completely fills the field of complex numbers that mathematicians refer to as the Riemann sphere.[139] Although the proof had to be worked out in an analytical way based on formulas, Douady insists that it was precisely the act of drawing that led to the mathematical insight: "And just at that moment, quite suddenly, as soon as we had made the drawing, we understood what was going on."[140]

In his handwritten chronology of the events, Douady emphasized the moment of recognition with a drawing interrupting the lines of text (fig. 34). Its shape, somewhere between an oval and a heart with loops turned inward, gives the impression of an eye gazing out between the lines. It can be regarded as a symbol for drawing as a thought process in mathematics.

A number of Douady and Hubbard's drawings for the "matings" are known. One of their preferred methods in the early 1980s was drawing on semitransparent foil. In figure 35, one can see how Douady traced different curving lines in red, blue, green, and violet. The "rabbit" is visible in red. The drawing is anchored in the trans-

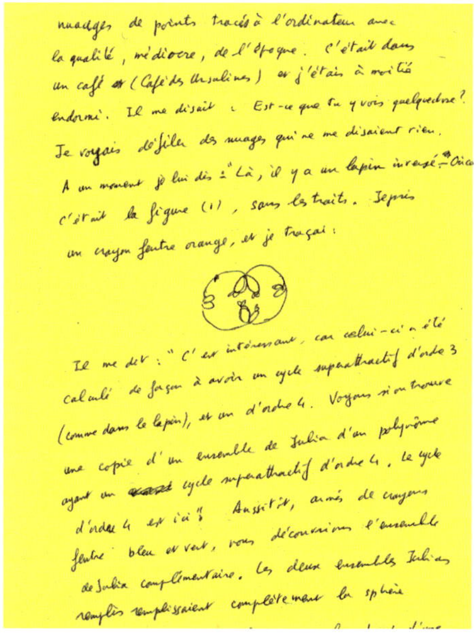
nuages de points tracés à l'ordinateur avec
la qualité, médiocre, de l'époque. C'était dans
un café ~~et~~ (Café des Ursulines) et j'étais à moitié
endormi. Il me disait : Est-ce que tu y vois quelquechose?
Je voyais défiler des nuages qui ne me disaient rien.
A un moment je lui dis : "Là, il y a un lapin inversé" Ouicar
c'était la figure (1), sans les traits. Je pris
un crayon feutre orange, et je traçai :

Il me dit : "C'est intéressant, car celui-ci a été
calculé de façon à avoir un cycle superattractif d'ordre 3
(comme dans le lapin), et un d'ordre 4. Voyons si on trouve
une copie d'un ensemble de Julia d'un polynôme
ayant un ~~[illegible]~~ cycle superattractif d'ordre 4, le cycle
d'ordre 4 est ici". Aussitôt, armés de crayons
feutre bleu et vert, nous découvrions l'ensemble
de Julia complémentaire. Les deux ensembles Julias
remplis remplissaient complètement la sphère

Fig. 34. Adrien Douady. Letter on the subject of the discovery of "fractal matings" (detail), December 20, 2004. Private Collection.

parent picture space by two slightly curving vertical lines. Vigorously curving loops and loosely distributed circles alternate with trial lines, some of which are struck through with hatching. These are traces of mathematical exploration that constitute a virtual allegory of the process of cognition, of its tentative searching, calculating, making of mistaken assumptions, and experiencing of reversal and reinforcement. Here, only by means of drawing, mathematical reflection could be

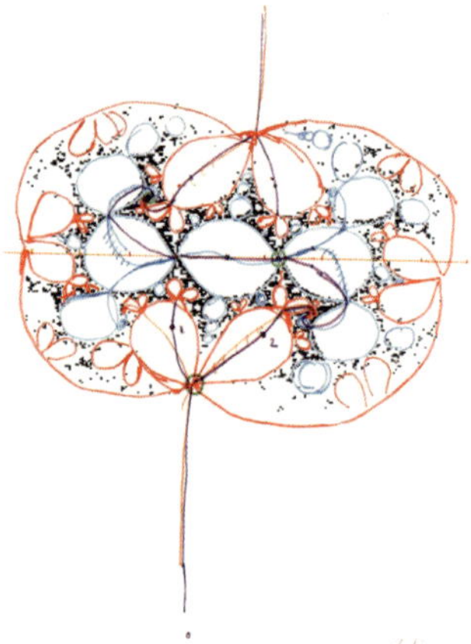

Fig. 35. Adrien Douady and John Hubbard. "Fractal matings," 1982. Markers on transparent foil and paper. Private Collection.

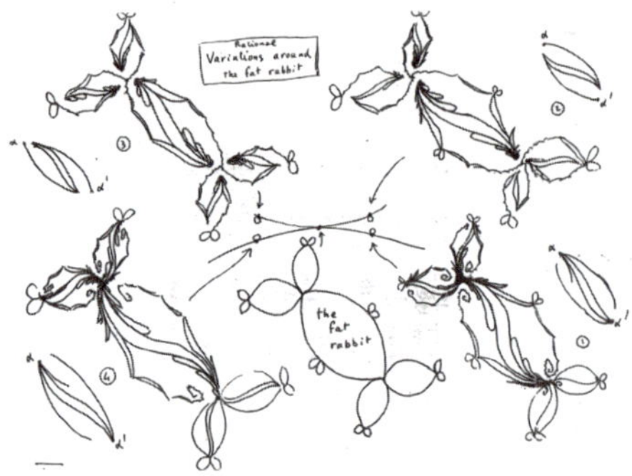

Fig. 36. Adrien Douady. Julia sets that cannot be visualized by computer. Drawn for the dissertation of Dan Sørensen, 1991. Courtesy Régine Douady.

seen as both controlling the digital space and becoming independent of it: With the use of foil as a transparent picture support, the visualized theory could be literally liberated from the mass of raw data. This drawing of course was not the proof, but it did provide evidence of what could later be confirmed. This is the core notion of the second possibility presented in this essay: that it is hand drawing that enables mathematical understanding in the realm of chaos, where serial pictures form sequences of similarities.

Analog and digital techniques are by no means mutually exclusive in fractal mathematics; they require and support one another. The drawing is not replaced by computer graphics; it enhances them and makes them intelligible. In fact, it is precisely owing to computer graphics that the pencil has become the mathematician's indispensable thinking tool when extracting a theoretical concept from the mass of visual data.

Douady, a skilled draftsman, produced a large number of other versions of his *lapin* (fig. 36). Made up of innumerable curves, curlicues, spirals, swirls, and snaking lines, his *lapins* explore geometrical relationships that can be proved mathematically, to be sure, but cannot be calculated and depicted by any computer. The center composite figure, labeled "the fat rabbit," is composed of five larger almond shapes and small, rounded buds. This basic figure is transposed into new configurations of serpentine lines in the surrounding drawings.[141] Douady's imaginative figures, drawn with great verve, show that research in this field of mathematics, first opened up by computer graphics, involves working with pictures even when the computer "blinds out," the formulas having become too complex for digital representation. At the time Douady drew them, the various curlicues he imagined were as yet nonexistent in a mathematical sense inasmuch as their mathematical properties were not accompanied by a rigorous proof. The proof would be the job of someone else: What Douady's drawing disclosed was a realm of possibility, a potential mathematical scenario.[142] Dan Sørensen, a doctoral student of Douady's, finally provided the proof of the scenario's existence—and of the efficacy of Douady's loopy imagination—in his dissertation.[143]

In Douady's sketches, the backward S-curve (see figs. 11 and 12) that the Bourbakists placed in the margins of their *Éléments de Mathématique* celebrated not only a return but an ecstatic revival. They also came to be symbolic of the factual force of drawings, for it was possible to prove mathematically the existence of the curlicue dynamic. Douady can thus be identified not only as a second-generation Bourbakist, but also as an exquisite Bourbakist draftsman.

Computer-graphic representations of fractals, especially those based on combinations of polynomials, are characterized by complex filigree structures around the edges. In order to study the relative position of their individual components, mathematicians draw “thought trees.” For example, as a combinatorial study, a drawing by Jean-Christophe Yoccoz from 1983 represents a small portion of the Mandelbrot set as mathematicians “see” it (fig. 37).[144] In his drawing, the calculation becomes a kind of hybrid between a tree and a molecular structure that fills the entire sheet. A basic function is seen at the bottom edge of the picture, a curving line reminiscent of a small hill. Rising up out of it is a gnarly trunk with straight and curved branches that sprout tiny twigs in every direction, an organic tangle whose leaves or fruits are numbers. Each of the tree’s branches is the expression of a calculating thought movement, and together they result in a complex dynamic of their own: The upper right is dominated by rhythmic lines curving to the left, repeating the movement of the exuberant formation

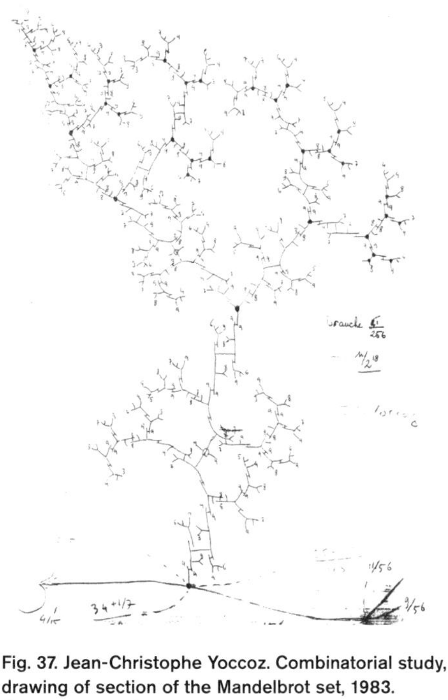

Fig. 37. Jean-Christophe Yoccoz. Combinatorial study, drawing of section of the Mandelbrot set, 1983. Courtesy Jean-Christophe Yoccoz.

in the upper left. Douady explained the combinatory practice as “tree drawing,” which represents one of the basic methods for working with polynomials and serves to tame the structures visualized in computer graphics.[145] Mathematical draftsmen thus become engineers of a computer-generated observation room; in drawings, digitally calculated abstract space is surveyed and virtually reconstructed. Computer graphics may have been required to disclose these realms to mathematics, but in scientific debate they had to be interpreted in drawings. Though the computer helped access a realm of pure thought in which the world of numbers might map itself seemingly unaided, drawing was essential for the return of the eye, which in Mandelbrot’s view was primarily manifested in digitalized mathematics.[146]

Drawings will always play an indispensable role in mathematical thinking at the theoretical stage. Yet the popular media are dominated by computer graphics that by and large mask the process of coming to understanding through drawing. To the media public, the presumably objective computer graphic is better at illustrating phenomena from the exact sciences. But the shape of the graphic has most likely been designed by a drawing hand, outside of public view, and only later presented in polished form.

Of course, it is possible to think in formulas—or, like Mandelbrot, in analogies—but arriving at understanding can also be a matter of fiddling with pencil and paper, making drawings that lead to a fruitful collaboration between the computer and the eye and thereby add to mathematical knowledge.[147]

Douady ended the article in which he presented his theory of mathematical *disegno* in 1982 (published, ironically, in the *Séminare N. Bourbaki*) with a highly technical demand: Construction on the foundations of the “chirurgy of dynamic systems” must continue.[148] His wording can be related to a humorous photomontage from his private archive (fig. 38), a spoof on a French leftist poster of the

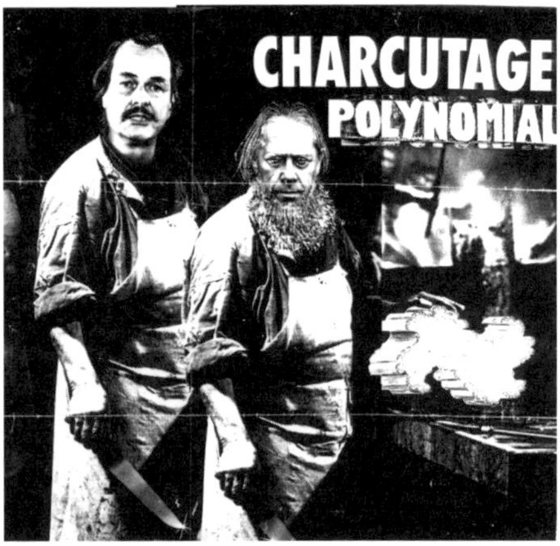

Fig. 38. Photo collage on paper with Adrien Douady (right) and John Hubbard, based on a poster of the PCF (Parti communiste française), 1986. Courtesy Régine Douady.

French Communist Party (PCF) in which John Hubbard and Adrien Douady have taken the places of the Gaullist politicians Jacques Chirac and Charles Pasqua. Douady's left hand rests on a three-dimensional drawing of a Julia set, which appears to hover slightly above a table (where in the original had been a map of France), while Hubbard is brandishing a butcher knife in his right hand. They are gazing at the viewer with determined expressions. The two mathematicians' intention is spelled out in large letters next to their heads: *Charcutage Polynomial* (Polynomial slaughter). This poster is a lively indication of the sparkling sense of humor for which Adrien Douady was widely known and appreciated. But there is a deeper meaning hidden in this joke: Had the two mathematicians not laid a hand and a pencil on the "slaughterhouse of the Julia sets," they would not have derived any information from the raw digital shapes of visualized mathematics. In this context, drawing became a mode of mathematical seeing: It was the linear that made the non-linear visible.

Material for Thought: Paper as a Threshold to the Mind; or, Paper *Poiesis*

A mathematician, on the other hand, has no material to work with but ideas.[149]
—Godfrey Harold Hardy

Anyway, I think with my knee.[150]
—Joseph Beuys

Experimental thinking, the cultural historian Thomas Macho has explained, does not take place solely in the mind: "Intellectual experiments are generally experiments on paper. We have to produce Flatland on our writing tables in order to be able to give an account of it."[151] Macho was of course referring to Edwin A. Abbott's 1884 novel in which the author describes a world where geometrical shapes play the main roles. Ninety years later, Otto E. Rössler's way of using pictures would provide an indication of how fundamentally important material evidence can be to experimental thinking in mathematics. With Rössler, our concluding case study is devoted to one of the most radically conceived and implemented collapsings of the boundaries between the world of thinking and the world of things that I encountered during my research in this field. We have already seen that in their mathematical thinking, Benoît Mandelbrot and Adrien Douady employed pictures in quite different ways. Rössler, in his work on the non-linear dynamics of complex systems, utilizes physical representations in his thinking in yet a third way. As with Douady, drawing is of utmost importance, but with Rössler, its emphasis expands dramatically from a focus on the visual side to its material qualities, which, in the manner of a feedback loop, directly influence the properties of the mathematical objects under study.

Rössler, a biochemist by training, is considered a pioneer in the field of "chaos theory" because of his mathematical works from the mid 1970s.[152] For the moment, the term "chaos" has deliberately been placed in quotation marks, as it conveys something very different to mathematicians than it does to laymen, for whom it is synonymous with "confusion" or "lawlessness." Chaotic systems form an order at the highest level. Accordingly, a more precise term is "deterministic chaos."[153] "Chaotic" means "not precisely predictable in the long term," to be sure, but by no means "arbitrary" or "unruly." Scientists working in this field have discovered forms of chaos as varied as they are precise. At the same time, they have suggested making chaos to a certain extent controllable; pictures of chaos are therefore always also "images of controllability." The field of fractals is linked to that of deterministic chaos. It can be thought of as a geometry of chaos. Thus, if deterministic chaos describes a temporal complexity, fractals deal with a spatial complexity.

In 1975, Rössler discovered an "attractor," a key concept in dynamics. Attractors can be described as spatially complex geometric shapes. In simple terms, they illustrate all the states a dynamic system can pass through over time. They are symbolic representations of long-term predictions. If they are chaotic, they are called "strange." The "Rössler attractor" is one such "strange" shape.[154] Rössler was deliberately searching for a "simplification" of the chaotic model discovered in 1963 by the meteorologist Edward Lorenz at MIT.[155] The Rössler model is extremely important in chaos research because it is reduced to the minimum requirements necessary to exhibit chaotic behavior and therefore keeps analysis of the dynamic system as simple as possible.

To explain how he first developed this shape, Rössler drew spiral-like tracks on two sheets of paper (fig. 39). He then proceeded to manipulate them in complex ways in three-dimensional space in an effort to understand them. By studying them one above the other but at an angle, he managed to "force in the paper"[156] a figure whose structure approaches the limit of what can be represented and imagined: an infinitely flat, two-dimensional thing described by the movement of a one-dimensional point that nevertheless occupies a three-dimensional space. It was only the manipulation of the two sheets of paper in various positions that gave him the sudden idea that the desired shape could only be realized by a dimensional leap. Once the line had arrived in the third dimension in its temporal development, it necessarily had to be catapulted back into two dimensions. Since it could be continually extended and retracted between the two sheets of paper arranged in space, it was indeed possible to return it to the second dimension.

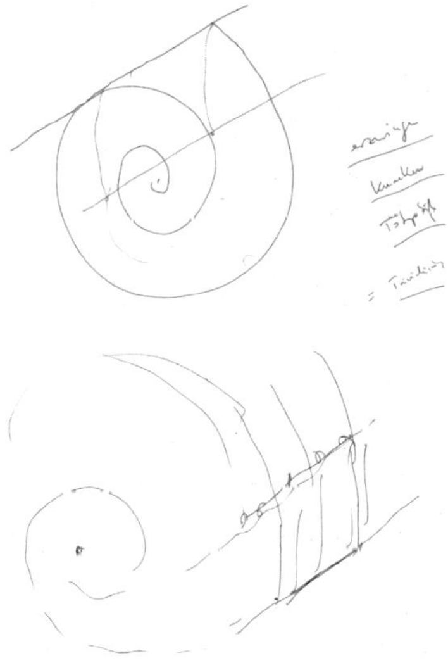

Fig. 39. Otto E. Rössler. Sketches to explain the discovery of the Rössler attractor, 2007. Pen on paper. Private Collection.

This account of his research methods and the circumstances in which he made his discovery in December 1975 dates from some thirty years after the events. Looking back, Rössler confirmed that it was in making the drawings and playing with them in space that he understood what he was after analytically; his subsequent development of the related formula—"the intrusion of a wedge of proof"[157]—was merely a mechanical act of realization. "[E]verything that one can imagine one can also draw," Rössler claimed, "and indeed much better, because in drawing one has what is essential. . . . It always begins with the pictures."[158] To emphasize this mechanism, Rössler noted a provocative triad that he had written on one of his sketch sheets (see fig. 39): "forcing—snapping—death bite" ("*erzwingen—knacken—Tötungsbiß*"). Retrospectively

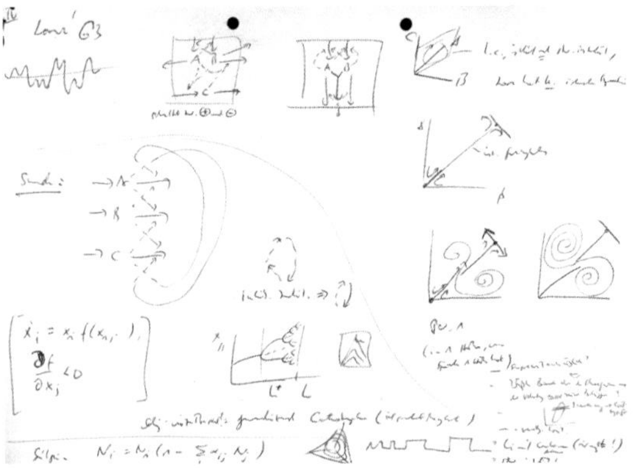

Fig. 40. Otto E. Rössler. Sketches on paper, 1975. Courtesy Otto E. Rössler.

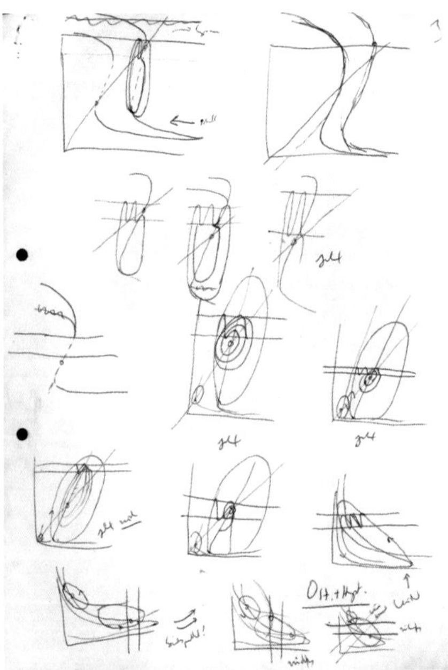

Fig. 41. Otto E. Rössler. Sketches on paper, 1975. Courtesy Otto E. Rössler.

surprised by the strong metaphors he had invented, he later commented that this "death bite" would certainly be a "friendly" or "natural" one, as achieved, for example, by the reflexes of a dog. It was as if the chaotic attractor had to be tamed in an actual tour de force, or in a process resembling "catching the prey." One could dismiss Rössler's description of his discovery through drawing as a fanciful projection after the fact were it not that recently discovered sketches document the process in detail (figs. 40–43). In scientific practice, preparatory sketches leading to a discovery are generally trashed, hardly worth preserving once the thinking process is finished. The preservation of this series of drawings is accordingly a historic stroke of luck. Dating to the period between November 23 and December 11, 1975, they provide eloquent testimony to Rössler's habit of thinking by drawing. One sees how he took up the basic shape of the Lorenz model—two interconnected spirals—in the first drawing (see fig. 40, right side) and then varied it step by step. He stretched the shape into an oval, deformed it into wavy shapes, and pressed it into the bottom corner of the coordinate system (see fig. 41).[159] Somewhat offhandedly, he wrote, "soft watches (Dalí)" (see fig. 42, left side) on another drawing. What was the name of the famous Surrealist painter doing among Rössler's working sketches? One has to know that in chemistry there are cyclical reactions known as "chemical clocks." On the left in figure 40, one sees a "diagram clock" that Rössler drew like an ear being struck by sound. Yet this equation does not produce a chaotic reaction, for in order to trigger one, the materiality of the image from chemistry—the clock—would need to be literally made "to melt," as in the famous Dali painting *The Persistence of Memory* (1931). This is explored in the other Rössler sketches here. Folding and stretching had already been identified in mathematics as characteristics of chaotic dynamics in the sixties; what was new was that by transferring the material malleability of the stable "chemical clock," a "chaos machine" might be created.[160] The idea that the material might be deformed was crucial to Rössler's subsequent discovery of the chaotic attractor through drawing. On the sheet with the "soft watches," there is also a sketch illustrating the step that immediately preceded Rössler's discovery (see fig. 42, upper right). It pictures the loops, now layered one behind the other in space, placed in a position to make possible the decisive step: the dimensional leap that Rössler characterized as the switch between two sheets

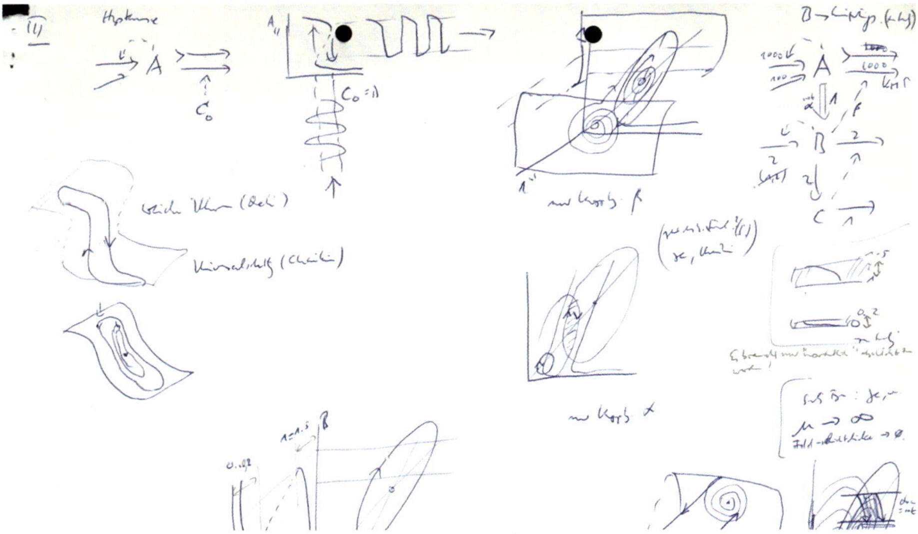

Fig. 42. Otto E. Rössler. Sketches on paper, 1975. Courtesy Otto E. Rössler.

of paper. In a small and seemingly unimposing sketch in the center of the last sheet, one sees the dimensional leap taking place in the more darkly hatched area between two broken lines (see fig. 43). Once Rössler had become aware of the ramifications of his discovery, he marked the small spiral drawing quite clearly with a green arrow—a kind of ex post facto exclamation point extolling the virtues of working out one's thinking on paper.

It was only after working out this series of simulations on paper that Rössler managed to come up with the equations for his "handmade" chaos model and to visualize them with the help of his analog computer (fig. 44).[161] The crucial preliminary role of his "paper models" and the relationship between the substance of his thinking and that which his hands could touch was highlighted still further when in 1977 he attempted to expand his model into the fourth dimension. To conceive of a new dimension at all, his hands had to first experience it intuitively. His experiments were now carried out with paper reshaped in space. Once again, a dimensional leap was required in the mathematical model, one that he could only grasp by rolling the paper.[162] Rössler published drawings of this paper model in the *Bulletin of Mathematical Biology* (fig. 45).[163] Thanks to

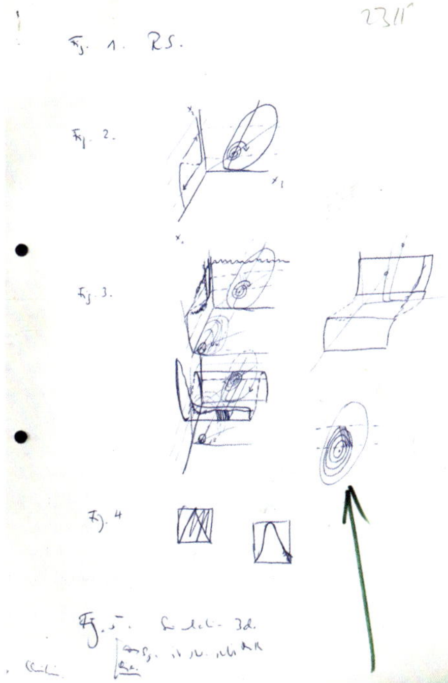

Fig. 43. Otto E. Rössler. Sketches on paper, 1975. Courtesy Otto E. Rössler.

the physical and tactile aids in his research, he was able to "pre-envision" what could only later be depicted in computer graphics with the help of formulas and algorithms. The computer alone would not have sufficed. Rössler's colleagues greeted his procedure

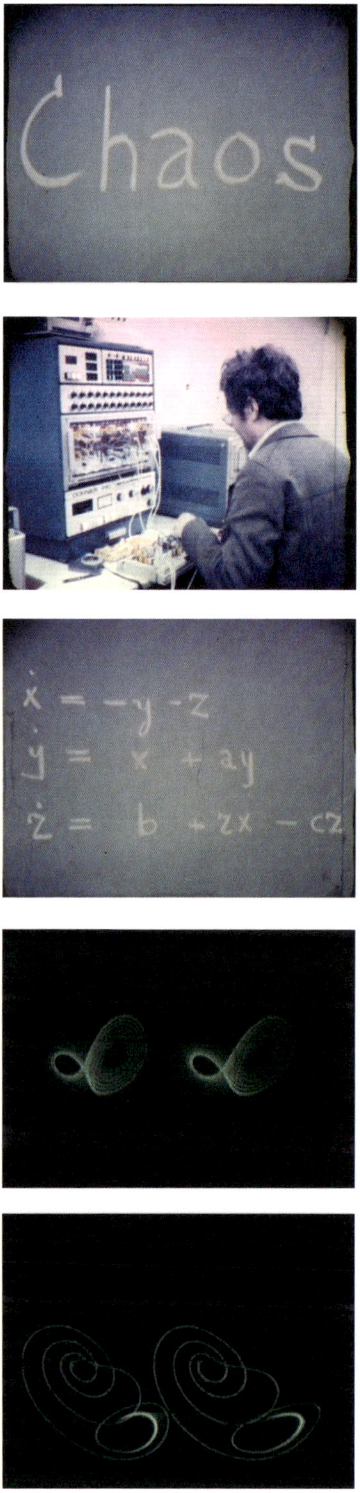

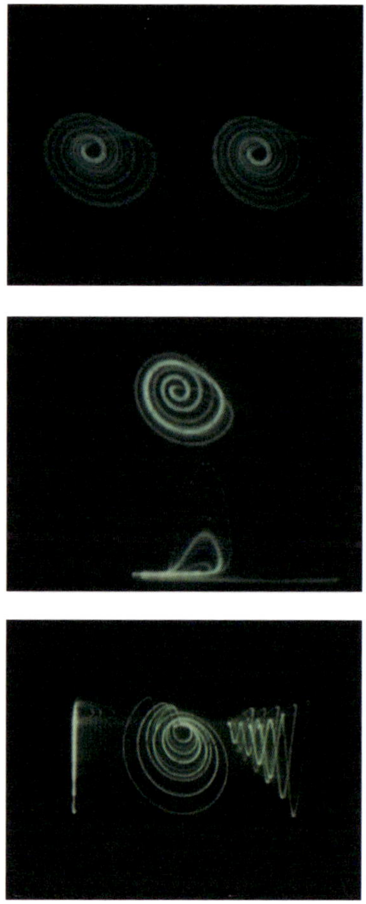

Fig. 44. Otto E. Rössler. *Chaos*, 1976–79. Computer-generated movie on analog computer. Stills. Courtesy Otto E. Rössler.

with amazed praise; as the mathematician Bruce Stewart put it: "What was brilliant about what Otto Rössler did was that he conceived of the shapes before the equations."[164]

Rössler's procedure can thus be seen as a radical expansion of Douady's position. Not only did the drawing trigger the sudden moment of understanding, the material substance—the paper model—seemed to sculpt the supposedly immaterial digital image itself. This opened up his thinking to the realm of possibility: "It is a mental experiment, but it is always real things that one manipulates."[165] The rolling of the paper and even the scratching of the pencil can be identified as "material springboards to

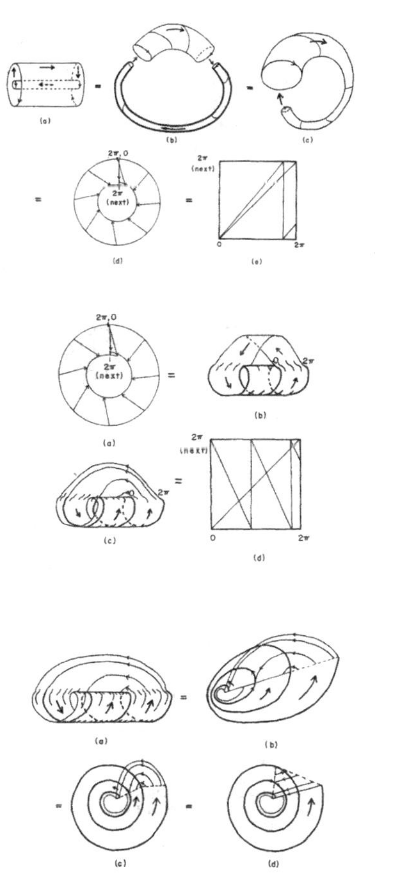

Fig. 45. Otto E. Rössler. "Rotation Maps, Equivalent Rotation Maps, and Paper Sheet Flow." From "Chaos in Abstract Kinetics: Two Prototypes," ***Bulletin of Mathematical Biology*** **39 (1977): 280, 282.**

thought" or "thinking thresholds," which influence the structure of creative thought, whether in art or mathematics. Rössler was, moreover, a sharp critic of digital pictures. Not only were they incapable of achieving a *return of the eye* into mathematics; strictly speaking, he insisted, they were not even "pictures": "A picture that is developed by way of algorithms is not truly a picture."[166]

Rössler describes the process of arriving at understanding by handling paper or drawing on it as one that has audible, even physical aspects: "If one makes a drawing on paper, or on two sheets of paper, at some point it snaps into place, it clicks. One has to force it, 'adjust' the shape so to speak. It then clicks audibly into place: It makes your ears pop."[167] A mental image takes shape through the gesture of drawing: It is a physical act just like any other physical experimentation, and as such it leads to understanding. Rössler quotes the behaviorist Konrad Lorenz's well-known statement that "Thinking is handling things."[168] In his reports, Rössler repeatedly relates how his own body becomes involved in his experiments. For example, he tells of a "nose perception" that he claims occurred as he was making his discovery. He had the feeling that a string kept winding around and around his nose. The figure of the spiral attractor taking shape in his imagination was virtually assaulting him. The researcher's own body here becomes a virtual "support" for the mathematical objects in defiance of the Western belief that the physical plays no part in "objective research."[169]

There is, at times, an air of the esoteric or even occult in Rössler's descriptions of his working method, one that is symptomatic of the intersubjectivity of mathematical picture worlds in general. To mathematicians, pictures that lead to understanding are often difficult to describe. The French mathematician Jacques Hadamard, who described his thinking in images as sequences of "undefined specks," provides a famous historical example. In his proof that there are prime numbers greater than 11, his first step read: "I consider all primes from 2 to 11, say 2, 3, 5, 7, 11." To his thinking, this was equivalent to saying: "I see a confused mass." Then in a final step, the closing solution: "I see a place somewhere between the confused mass and the first point."[170] One has to imagine the intervening mental steps—that is to say, the path from "confused mass" to the previous point—as moving images or a dynamic sequences of pictures. It is not known whether or when Hadamard made drawings. But his insight that mathematical thinking occurs in visual and not always clearly definable images unites him with Rössler.[171] In Rössler's case, however, one has to add that these "mental images" can no longer be separated from the material world. As a process, thought itself becomes material,

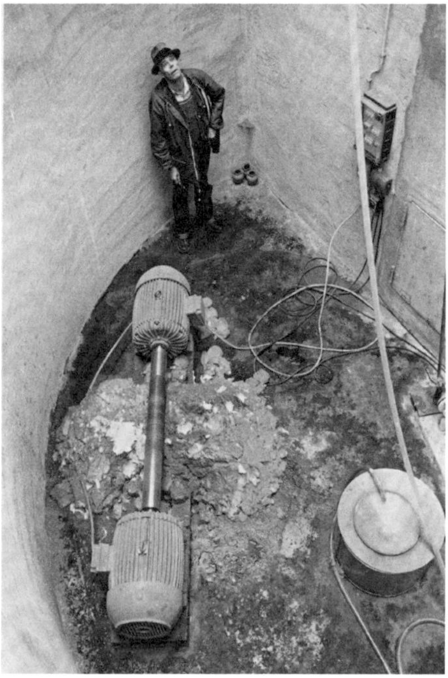

Fig. 46. Joseph Beuys. Installation view of *Honey Pump in the Workplace (Honigpumpe am Arbeitsplatz)*, documenta 6, Fridericianum Kassel, June–October 1977. Photograph. Dieter Schwerdtle / © documenta Archiv. © 2012 Artists Rights Society (ARS), New York/VG Bild-Kunst, Bonn.

and this materiality of thinking can be seen as a habit that does not necessarily have to be conscious: "You just introduced the term *haptic*—that was a new formulation for me, but you are correct. It is only new to me as a name, not as experience."[172]

In this, Rössler can be said to agree with Friedrich Nietzsche that "our writing tools influence our thinking."[173] Whereas Nietzsche loved to "think while writing,"[174] to Rössler there is something erotic about pictures: "Real pictures are always erotic pictures as well. Topology or geometry is like looking from all sides as though in outer space. That is all completely tactile. Without touching and without kneading it is impossible to imagine. Materiality does matter: Essentially mathematics is nothing more than pottery."[175]

One has to resist confusing Rössler's creative process as a mathematician with that of Prometheus, who as a skilled sculptor is said to have fashioned mankind out of clay. Rössler's approach is more reminiscent of the "organic machines" designed in the mid-seventies by the artist Joseph Beuys, in which materiality was basically perceived from the point of view of its energy potential. In *Honey Pump in the Workplace*, created by Beuys for documenta 6 in 1977, the artist pumped honey, conceived of as a source of energy, through a system of plastic tubing (fig. 46). With this "model of man," Beuys designed a "technology . . . that is first of all, in fact, an internal one, that is a conceptually creative one in mankind himself."[176] Just as the circulating honey can be understood as the circulation of blood of such an "organic machine,"[177] Rössler's rolling of paper sheets was a way of designing chaos by thinking in material terms. Prometheus molded man out of clay so as to breathe life into him and give him the ability to reason. Rössler molded paper as a way of opening up to his thinking new realms of possibility.

Epilogue: Interfaces

The mathematical mind has always seen itself confronted with the problem of generating sense out of sequences of similarities. Here we have been able to distinguish three basic ways of pictorial thinking by examining the working materials of three different scientists. Benoît Mandelbrot elevated analogy to a guiding principle and saw new pictures of the world taking shape in oscillating similarities, while his islands pointed to the double-edged, imaginative, and seductive power of the visual and identified drawing as means of authentication. For Adrien Douady, the drawing of a line as a kind of mental sword through the drift of dots in a series of similar computer graphics led to a breakthrough in understanding. To Otto E. Rössler, the experimental manipulation of things was equivalent to thinking—and all the images in the various configurations of his sheets of paper were one and the same picture. His research method attested to the dignity of the single picture as such. Rolling his sheet of paper while pondering four-dimensional chaos, he was not merely rolling paper. His thinking was rolling, or alternatively, he was thinking the rolling, or perhaps even better, the rolling was thinking him. As Maurice Merleau-Ponty was also aware, "One has to understand that things own us, not we things."[178]

Creative processes in art and the methods of the sciences both suggest that mental and physical activities are interdependent and influence each other. The practice of "thinking with one's hand," the subject of the present book and the related exhibition, coincides with Beuys's thinking "with my knee"[179]—both are suggestive of a performative model of thought. It is simply mistaken to assume that there is an autonomous mental realm independent of the world of sense impressions, one in which ideas and concepts can be pursued without reference to the material world—whether in the practice of art or mathematics.

When trying to categorize more closely the *evidence*—or *traces*—that mathematical objects leave behind in the world, as Jean-Pierre Changeux proposed at the very beginning of this essay, it is only possible to conceive of them as interfaces between the world of mathematical thought and visible reality. It is their very double allegiance that gives them their potency and is the basis for their unique dynamic: They do not dissipate somewhere between the two spheres but leave behind traces, not only on paper but in thinking itself.

1 Jean-Pierre Changeux and Alain Connes, *Conversations on Mind, Matter, and Mathematics* (Princeton: Princeton University Press, 1995). The original title of their book, *Matière à Pensée*, evokes the double meaning of "materials to think about" and "the materiality of thought." The present essay builds on previous of my publications, starting with my dissertation, "Die Form des Chaos. Bild und Erkenntnis in der komplexen Dynamik und der fraktalen Geometrie" (forthcoming, 2013, from Fink Publishers) as well as "Form und Farbe digitaler Mathematik: Vom Zusammenspiel von zeichnender Hand und Computer in fraktalen Bildwelten," *Bildwelten des Wissens, Kunsthistorisches Jahrbuch für Bildkritik* 3 (2005): 18–31, and "'I Look, Look, Look, and Play with Many Pictures,' Zur Bilderfrage in Benoît Mandelbrots Werk," in *Verwandte Bilder. Die Fragen der Bildwissenschaft*, ed. Ingeborg Reichle, Steffen Siegel, and Achim Spelten (Berlin: Kadmos, 2007), 297–320.

2 Changeux and Connes, *Conversations on Mind, Matter, and Mathematics*, 44. Italics in the original. While their publication contains some illustrations, they do not go into particulars.

3 Benoît Mandelbrot, *Fractals and Chaos: The Mandelbrot Set and Beyond* (New York: Springer, 2004), 34. Italics in the original.

4 Their existence was known, nevertheless, as he mentions them in Mandelbrot, "People and Events Behind the 'Science of Fractal Images,'" in *The Science of Fractal Images* (New York: Springer-Verlag, 1988), 8. One of the films was titled "Competing with the Lord on the First Day of Creation."

5 Benoît Mandelbrot, "Intermittent Turbulence and Fractal Dimension: Kurtosis and the Spectral Exponent 5/3+B," in *Turbulence and Navier Stokes Equations* (Orsay, 1975), ed. Roger Temam (New York: Springer, 1976), 121.

6 H. G. Wells, *The Island of Dr. Moreau* (Garden City, NY: Garden City Publishing Co., 1896); see also R. M. P. Bozzetto and Russell Taylor, "Moreau's Tragi-Farcical Island," *Science Fiction Studies* 20 (1993): 34–44. See also Günter M. Ziegler, "Countdown. Über Mathematik und Science Fiction," in *Gegenworte. Hefte für den Disput über Wissen. Bilder (in) der Wissenschaft* 20 (2008): 73–82.

7 For this motif, see Horst Bredekamp, "Der Mensch als 'zweiter Gott.' Motive der Wiederkehr eines kunsttheoretischen Topos im Zeitalter der Bildsimulation," in *Bilder bewegen. Von der Kunstkammer zum Endspiel*, ed. Jörg Probst (Berlin: Wagenbach, 2007).

8 See Margarete Pratschke, "Charles and Ray Eames' 'Powers of Ten'—Die künstlerische Bildfindung des Atoms zwischen spielerischem Entwurf und wissenschaftlicher Affirmation," in *Atombilder. Ikonografien des Atoms in Wissenschaft und Öffentlichkeit des 20. Jahrhunderts*, ed. Charlotte Bigg and Jochen Hennig (Göttingen: Wallstein-Verlag, 2009), 21–30.

9 Benoît Mandelbrot, interview by the author, Cambridge, MA, April 2008.

10 Benoît Mandelbrot, *The Fractal Geometry of Nature* (New York: Freeman, 1982), 21.

11 Magdolna Hargittai and István Hargittai, *Candid Science Four* (London: Imperial College Press, 2004),

497. Other encomiums were similarly offered: As a commencement speaker at Syracuse University, which awarded him an honorary doctorate in Science and Engineering in 1986, he was lauded for "unraveling the secrets of the universe"; when Boston University made him a Doctor of Science in 1987, it was in recognition of "scientific, mathematical, and aesthetic insights by which you have brought us closer to the character of the universe"; see http://www.math.yale.edu/mandelbrot (accessed on December 10, 2011).

12 For photographic portraits of scientists and their relation to popularization, see Daniel Jacobi and Bernard Schiele, "Scientific Imagery and Popularized Imagery: Differences and Similarities in the Photographic Portraits of Scientists," *Social Studies of Science* 19 (1989): 731–53. For the tradition of scientists' portraits deriving from the nineteenth century, see Gabriele Werner, "Das Bild vom Wissenschaftler—Wissenschaft im Bild. Zur Repräsentation von Wissen und Autorität im Portrait am Ende des 19. Jahrhunderts," *kunsttexte.de* 1 (2001): 1–11.

13 For his broken Dimension D—the "Hausdorff Dimension," an important feature of fractals—Mandelbrot adapted a definition formulated in Felix Hausdorff's "Dimension und äusseres Mass," *Mathematische Annalen* 79 (1919): 157–79.

14 "LBLGRAPH" (or Labelgraph) was the name of a graphic program at IBM; see Sigmund Handelman, "A High-Resolution Computer Graphics System," *IBM Systems Journal* 19 (1980): 361.

15 Benoît Mandelbrot, "People and Events Behind the 'Science of Fractal Images,'" in *The Science of Fractal Images* (New York: Springer, 1988), 8–10. See Alain Fournier, Don Fussell, and Loren Carpenter, "Computer Rendering of Stochastic Models," *Communications of the ACM* 25 (1982): 371–84.

16 Benoît Mandelbrot, *The Fractal Geometry of Nature*, 21.

17 See Benoît Mandelbrot, "Les inattendus des fractales," *pour la science* 234 (1997): 10–11.

18 For a mathematical critique, see Steven G. Krantz, "Fractal Geometry," *The Mathematical Intelligencer* 11 (1989): 12–16.

19 Gottfried Boehm, "Die Wiederkehr der Bilder," in *Was ist ein Bild?*, ed. Gottfried Boehm (Munich: Fink, 1994), 35. Boehm here refers in general to "electronic simulation techniques." For fundamental contributions to the critical art-historical discussion, see Karl Clausberg, "Feigenbaum und Mandelbrot. Neue Symmetrien zwischen Kunst und Naturwissenschaften," *Kunstforum International* 85 (1986): 86–93; Karl Clausberg, "Am Ende Kunstgeschichte? Künstliche Wirklichkeiten aus dem Computer," in *Kunstgeschichte—aber wie? Zehn Themen und Beispiele*, ed. Clemens Fruh, Raphael Rosenberg, and Hans-Peter Rosinski (Berlin: Reimer, 1989), 259–93; and Horst Bredekamp, "Mimesis, grundlos," *Kunstforum International* 114 (1991).

20 See the exemplary Jean Baudrillard, *Die Agonie des Realen* (Berlin: Merve, 1978); Jean-François Lyotard, *La Condition postmoderne: rapport sur le savoir* (Paris: Les Éditions de Minuit, 1979); and Baudrillard, *Subjekt und Objekt: fraktal. Vortrag im Kunstmuseum Bern* (Bern: Wabern, 1986). For a reaction to fractals and chaos in the context of postmodernism, see Vilém Flusser, "Curie's Children: Vilém Flusser on an Unspeakable Future," *Artforum* 19 (1990): 22–23; Vivian Sobchack, "A Theory of Everything: Meditations on Total Chaos," *Artforum* 19 (1990): 48–55; Michael Speaks, "Chaos, Simulation, and Corporate Culture," *Mississippi Review* 17 (1989): 159–76; Richard Wright, "Art and Science in Chaos: Contested Readings of Scientific Visualizations," in *FutureNatural: Nature, Science, Culture*, ed. George Robertson (London: Routledge, 1996); Horst Bredekamp, "Der simulierte Benjamin. Mittelalterliche Bemerkungen zu seiner Aktualität," in *Frankfurter Schule und Kunstgeschichte*, ed. Andreas Berndt, et al (Berlin: Reimer, 1992), 117–40; and Horst Bredekamp, "Das Bild als Leitbild. Gedanken zur Überwindung des Anikonismus," in *LogIcons: Bilder zwischen Theorie und Anschauung* (Berlin: Sigma, 1997), 233–36.

21 This information is based on a conversation with Aliette Mandelbrot on December 4, 2011, in Cambridge, MA.

22 Benoît Mandelbrot, "Fractals and the Rebirth of Experimental Mathematics," in *Fractals for the Classroom: Part One: Introduction to Fractals and Chaos*, ed. Hartmut Jürgens, Heinz-Otto Peitgen, and Dietmar Saupe (New York: Springer, 1992), 4.

23 Georg Cantor, quoted from Jean-Luc Chabert, "Un demi-siècle de fractales: 1870–1920," *Historia Mathematica* 17 (1990): 348.

24 Charles Hermite, letter from May 20, 1893, in Benjamin Baillaud and Henry Bourget, eds., *Correspondance d'Hermite et de Stieltjes*, vol. 2 (Paris: Gauthier-Villars, 1905), 318.

25 See Klaus Thomas Volkert, *Die Krise der Anschauung: Eine Studie zu formalen und heuristischen Verfahren in der Mathematik seit 1850* (Göttingen: Vandenhoeck & Ruprecht, 1986); Herbert Mehrtens, *Moderne—Sprache—Mathematik: Eine Geschichte des Streits um die Grundlagen der Disziplin und des Subjekts formaler Systeme* (Frankfurt am Main: Suhrkamp, 1990); and Bettina Heintz, *Die Herrschaft der Regel. Zur Grundlagengeschichte des Computers* (Frankfurt am Main: Campus Verlag, 1993).

26 Paul Du Bois-Reymond, "Versuch einer Classification der willkürlichen Functionen reeller Argumente nach ihren Aenderungen in den kleinsten Intervallen," *Journal für die reine und angewandte Mathematik* 79 (1875): 29 note.

27 Henri Poincaré, *Science et Méthode* (Paris: Flammarion, 1908), 132.

28 Benoît Mandelbrot, "Des monstres de Cantor et de Peano à la géométrie fractale de la nature," in Jean A. Dieudonné et al., eds., *Penser les mathématiques* (Paris: Seuil, 1982), 233; for the aspect of "beauty," see also Mandelbrot, "Les fractals, les monstres et la beauté," *Le débat: histoire, politique, société* 24 (1983): 54–72. For the development of the terminology of the "monsters," see Klaus Thomas Volkert, *Die Krise der Anschauung,* 99 ff.

29 Angela Fischel, *Natur im Bild: Zeichnung und Naturerkenntnis bei Conrad Gessner und Ulisse Aldrovandi* (Berlin: Mann, 2009), 118.

30 Georg Cantor, quoted from Chabert, "Un demi-siècle de fractales: 1870–1920," 348.

31 Mandelbrot, "Des monstres de Cantor et de Peano à

la géométrie fractale de la nature," 226.

32 Bernhard Siegert, *Passage des Digitalen. Zeichenpraktiken der neuzeitlichen Wissenschaften 1500–1900* (Berlin: Brinkmann & Bose, 2003), 313, 318.

33 See, for example, David Hilbert, "Über die stetige Abbildung einer Linie auf ein Flächenstück," *Mathematische Annalen* 38 (1891): 459–60.

34 See Volkert, *Die Krise der Anschauung*, 90.

35 Hermann Weyl, "Über die neue Grundlagenkrise der Mathematik," in *Hermann Weyl. Gesammelte Abhandlungen*, ed. Komaravolu Chandrasekharan, vol. 2 (Berlin: Springer, 1968), 143.

36 See David Hilbert, *Grundlagen der Geometrie. Festschrift zur Feier der Enthüllung des Gauss-Weber-Denkmals in Göttingen* (Leipzig: B. G. Teuber, 1899), 459–60.

37 For Hilbert's radical axiomatic program, see Herbert, Mehrtens, *Moderne—Sprache—Mathematik,* 108–42.

38 Ibid., 113.

39 See Heintz, *Die Herrschaft der Regel. Zur Grundlagengeschichte des Computers*, 16–34. See also Albert Einstein's corresponding description of the axiomatic method in Albert Einstein, *Geometrie und Erfahrung. Erweiterte Fassung des Festvortrages gehalten an der Preußischen Akademie der Wissenschaften zu Berlin am 27. Januar 1921* (Berlin: Springer, 1921).

40 Konrad Knopp, *Mathematik und Kultur* (Berlin: W. de Gruyter, 1985), 22.

41 See Paul R. Halmos, "Nicolas Bourbaki," *Scientific American* 196 (1957): 89; also Jean A. Dieudonné, "The Work of Nicholas Bourbaki," *The American Mathematical Monthly* 77 (1970): 134 note. The name of the general concerned was Charles Denis Bourbaki.

42 Nicolas Bourbaki, "Die Architektur der Mathematik," in *Mathematiker über Mathematik*, ed. Michael Otte (Berlin: Springer, 1974), 158.

43 Ibid., 144.

44 According to Pierre Cartier, quoted in Marjorie Senechal, "The Continuing Silence of Bourbaki: An Interview with Pierre Cartier," *The Mathematical Intelligencer* 20 (1998): 27.

45 Bourbaki, "Die Architektur der Mathematik," 151.

46 Armand Borel, *Mathematik: Kunst und Wissenschaft* (Munich: Die Stiftung, 1982), 33.

47 Bourbaki, "Die Architektur der Mathematik," 158 ff.

48 See Maurice Mashaal, *Bourbaki. Une société secrète de mathématiciens* (Paris: Belin, Pour la science, 2002), 56.

49 Horst Bredekamp, "Die Zeichnende Denkkraft. Überlegungen zur Bildkunst der Naturwissenschaften," in *Einbildungen*, ed. Jörg Huber (Zurich: Springer, 2005), 163–65.

50 For the *figura serpentinata* of the sixteenth century and its art-theoretical reinterpretation by Hogarth, see Peter Gerlach, "Zur zeichnerischen Simulation von Natur und natürlicher Lebendigkeit," *Zeitschrift für Ästhetik und Allgemeine Kunstwissenschaft* 34 (1989): 243–79.

51 See Régine Bonnefoit, *Die Linientheorien von Paul Klee* (Petersberg: Imhof, 2009), 9, 43–47. Graphical and theoretical investigations of the expressivity of lines are no exception in the early twentieth century (cf. psychology, philosophy, graphology, choreography). For a discussion of the serpentine line as a time-transcending symbol of visual thinking and the movements of the mind, see Horst Bredekamp, "Die Unüberschreitbarkeit der Schlangenlinie," in *Minimal concept: zeichenhafte Sprachen im Raum*, ed. Christian Schneegass (Dresden: Verlag der Kunst, 2001), 205–208.

52 David Hilbert, "Mathematische Probleme," in *Die Hilbertschen Probleme*, ed. P. S. Alexandrov (Leipzig: Akademische Verlagsges, Geest & Portig, 1983), 27

53 In this spirit, the Old Testament refers to the symbol of a copper snake, looking at which would save the life of the snake-bitten (cf. Numbers 21:9). In this narrative, the mimetic reproduction of the threatening phenomenon was credited with the power to defeat itself.

54 See Allyn Jackson, "Comme Appelé du Néant—As If Summoned from the Void: The Life of Alexandre Grothendieck, II," *Notices of the AMS* 51 (2004): 1199. He called his "picture theory" "dessins d'enfants," with pictures reminiscent of children's line drawings becoming the objects of investigation.

55 Benoît Mandelbrot, see interview in this volume, p. 155.

56 Mandelbrot, interview, April 2008.

57 Benoît Mandelbrot, see interview in this volume, p. 154

58 Benoît Mandelbrot, *Fractals and Chaos: The Mandelbrot Set and Beyond* (New York: Springer, 2004), 3.

59 Benoît Mandelbrot, see interview in this volume, p. 154.

60 Ibid.

61 Benoît Mandelbrot, "How Long Is the Coast of Britain? Statistical Self-Similarity and Fractional Dimension," *Science* 156 (1967): 636–38.

62 Lewis Fry Richardson, "The Problem of Contiguity: An Appendix to Statistics of Deadly Quarrels," in *General Systems: Yearbook of the Society for the Advancement of General Systems Theory* 6 (1961): 168. Like Mandelbrot, Richardson was an interdisciplinary border crosser, involved in meteorology, mathematics, geophysics, statistics, and even peace studies.

63 Quotes from Mandelbrot, "Les inattendus des fractales," 10–11, and Mandelbrot quoted from Monte Davis, "Profile of Benoît B. Mandelbrot," *Omni Magazine* (February 5, 1984): 4.

64 Peter Glaser, "Das Kolumbus-Gefühl. Entdeckungen in einer virtuellen Welt," in *Computerkultur oder The Beauty of Bit & Byte*, ed. Michael Weisser (Bremen: TMS, 1989), 19–40. Comparable descriptions—"a new world to conquer"—are found in connection with microscopy; see Stefan Ditzen, "Der Satyr auf dem Larvenrücken. Zum Verhältnis von instrumentellem Sehen und Bildtraditionen," in *Konstruierte Sichtbarkeiten. Wissenschafts- und Technikbilder seit der Frühen Neuzeit*, ed. Martina Hessler (Munich: W. Fink, 2006).

65 Benoît Mandelbrot, *Les objets fractals: forme, hasard et dimension* (Paris: Flammarion, 1975), 116.

66 Benoît Mandelbrot, "On the Geometry of Homogeneous Turbulence, with Stress on the Fractal Dimension of the Iso-Surfaces of Scalars," *Journal of Fluid Mechanics* 72 (1975): 405.

67 For chance methods in the creation of twentieth-century art, see Bernhard Holeczek and Lida von Mengden, eds., *Zufall als Prinzip: Spielwelt, Methode und System in der Kunst des 20. Jahrhunderts*, exh.

cat. (Ludwigshafen am Rhein: Wilhelm-Hack-Museum, 1992). For chance as a principle of the artistic avant-garde, see Christian Janecke, *Kunst und Zufall. Analyse und Bedeutung* (Nuremberg: Verlag für Moderne Kunst, 1995); Horst Bredekamp, "John Cage and the Principle of Chance," in *Music and the Aesthetics of Modernity*, ed. Karol Berger and Anthony Newcomb (Cambridge, MA: Harvard University Press, 2005), 99–107. Herbert Molderings, *Duchamp and the Aesthetics of Chance: Art as Experiment* (New York: Columbia University Press, 2010). For chance pictures as "picture games of nature," see Horst Bredekamp, *Theorie des Bildakts: Über das Lebensrecht des Bildes* (Frankfurt am Main: Suhrkamp, 2010), 317–23.

68 Giorgio Vasari, *Vasari on Technique*, trans. Louisa S. Maclehose (London: Dent, 1907), 212. In later centuries as well, the relationship between blot and picture existed at the interface between art and science, serving as a field for experimentation for the testing and expansion of concepts like truth and objectivity. See Friedrich Weltzien, *Fleck, das Bild der Selbsttätigkeit: Justinus Kerner und die Klecksografie als Experimentelle Bildpraxis zwischen Ästhetik und Naturwissenschaft* (Göttingen: Vandenhoeck & Ruprecht, 2011). In relation to Mandelbrot's interpretation of specks, another cultural technique comes to mind, this time at the interface between psychosomatics and hermeneutics: Rorschach blots, the technique developed by Hermann Rorschach in which ink blots are used as a diagnostic tool for the psychological testing of personality; see Hermann Rorschach, *Psychodiagnostik. Methodik und Ergebnisse eines wahrnehmungsdiagnostischen Experiments (Deutenlassen von Zufallsformen)*, ed. Walter Morgenthaler (Bern: Hans Huber, 1972). According to Peter Galison, Rorschach tests became a "symbolic technology" that made earlier psychological and philosophical debates about the distinction between "seeing" and "seeing as" superfluous; see Peter Galison, "Image of Self," in *Things That Talk: Object Lessons from Art and Science*, ed. Lorraine Daston (New York: Zone Books, 2004), 276, 291–93.

69 Vasari, *Vasari on Technique*, 212.

70 Alexander Cozens, "An Essay to Facilitate the Invention of Landskips," quoted in Werner Busch, *Das sentimentalische Bild: die Krise der Kunst im 18. Jahrhundert und die Geburt der Moderne* (Munich: Beck, 1993), 337.

71 Alexander Cozens, "New Method of Assisting the Invention in Drawing Original Compositions of Landscape," quoted from Busch, *Das sentimentalische Bild*, 348. Cozens thought of his manual as an expansion of the method suggested by Leonardo da Vinci in his Treatise on Painting through which forms encountered by chance in nature are used as sources of inspiration for motifs in painting. In contrast to Leonardo, Cozens preferred to produce the forms deliberately, "making those imperfect Forms on Purpose, and with some degree of Design, which otherwise must be catch'd from Accidents"; Cozens, "An Essay to Facilitate the Invention of Landskips," 1, quoted from Kim Sloan, "A New Chronology for Alexander Cozens Part II: 1759–86," *The Burlington Magazine* 127 (June 1985): 356, illus. 7.

72 Busch, *Das sentimentalische Bild*, 346.

73 Holeczek and von Mengden, eds., *Zufall als Prinzip*, 17.

74 Busch, *Das sentimentalische Bild*, 348.

75 Hans-Jörg Rheinberger, *Experimentalsysteme und epistemische Dinge: eine Geschichte der Proteinsynthese im Reagenzglas* (Frankfurt am Main: Suhrkamp, 2006), 27, 128. This corresponds to Rheinberger's description of the peculiarity of an "epistemic thing"; see ibid., 27–34.

76 Benoît Mandelbrot, "New Methods in Statistical Economics," *The Journal of Political Economy* 71 (1963): 435.

77 William Feller, *An Introduction to Probability Theory and Its Applications*, vol. 1 (New York: Wiley, 1957), 83–85. This image only appears in the 2nd edition.

78 Mandelbrot, "New Methods in Statistical Economics," 435.

79 Mandelbrot, *Les objets fractals*, 55; Mandelbrot, *Fractals*, 24; Mandelbrot, *The Fractal Geometry of Nature*, 257. In economics and signal transmission: J. M. Berger and Benoît Mandelbrot, "A New Model for Error Clustering in Telephone Circuits," *IBM Journal of Research and Development* 7 (1963): 225; Mandelbrot, "Self-Similar Error Clusters in Communication Systems and the Concept of Conditional Stationarity," *IEEE Transaction on Communications Technology* 13 (1965): 73. Further, in relation to Louis Bachelier, in Mandelbrot, "Le syndrome de la variance infinie et ses rapports avec la discontinuité des prix," *Economie Appliquée* 26 (1973): 325.

80 See Edward R. Tufte, *The Visual Display of Quantitative Information* (Cheshire, CT: Graphics Press, 2002), 32. See also Howard Wainer, "Graphical Visions from William Playfair to John Tukey," *Statistical Science* 5 (August 1990): 340–46; and A. D. Biderman, "The Playfair Enigma: The Development of the Schematic Representation of Statistics," *Information Design* 6 (1990): 3–25.

81 In the second half of the eighteenth century, geology developed as a scientific discipline with its own pictorial tradition; see Martin Rudwick, "The Emergence of Visual Language for Geological Science 1760–1840," *History of Science* 14 (1976): 149–95; also Rudwick, *Bursting the Limits of Time: The Reconstruction of Geohistory in the Age of Revolution* (Chicago: University of Chicago Press, 2005).

82 One of the chapters in Ruskin's major work, *Modern Painters*, from 1856, was titled "The Sculpture of Mountains." Ruskin was influenced by Charles Lyell, the founder of modern geology, who in the years 1830–33 had questioned the notion of biblical creation with his publication of *Principles of Geology*; see Moritz Wullen, ed., *Natur als Vision: Meisterwerke der englischen Präraffaeliten* (Berlin: SMB-DuMont, 2004), 43, 54. Like Mandelbrot later, Ruskin was of the opinion that mountains are built up scale-invariantly and that therefore the structure of an entire mountain is recognizable in the smallest rocks, a view he found expressed in similar fashion in the "nature vision" paintings of the Pre-Raphaelites, whom he admired and promoted. Ruskin's thesis claimed that "the laws of geological processes . . . can be visualized in a single piece of rock just as in an entire mountain"; ibid., 46.

83 Compare the reception of the Feynman Diagrams, adapted to visual conventions. See David Kaiser, "Stick-Figure Realism: Conventions, Reifications, and the Persistence of Feynman Diagrams, 1948–1964," *Representations* 70 (2000): 49–86.

84 According to the art historian Gottfried Boehm, scientific pictures are *deictic*, which means that their chief function is "to show." Given their instrumental function, such aesthetic criteria as wealth of allusion, metaphorical quality, visual density, and self-reference hardly come into play; see Gottfried Boehm,"Zwischen Auge und Hand. Bilder als Instrumente der Erkenntnis," in *Mit dem Auge denken. Strategien der Sichtbarmachung in wissenschaftlichen und virtuellen Welten*, ed. Bettina Heintz and Jörg Huber (Vienna: Springer, 2001), 53.

85 Margit Rosen, "Die Maschinen sind angekommen. Die [Neuen] Tendenzen—visuelle Forschung und Computer," *bit international. [Nove] tendencije—Computer und visuelle Forschung. Zagreb 1961–1973*, ed. Margit Rosen and Peter Weibel (Graz: Neue Galerie, 2007), 39.

86 See Karin Gludovatz, "Malerische Worte. Die Künstlersignatur als Schrift-Bild," in *Schrift: Kulturtechnik zwischen Auge, Hand und Maschine*, ed. Gernot Grube, Werner Kogge, and Sybille Krämer (Munich: W. Fink, 2005), 313. For signatures in the sciences, see Alfred Nordmann, "Vor-Schrift—Signaturen der Visualisierungskunst," in *Ästhetik der Wissenschaft. Interdisziplinärer Diskurs über das Gestalten und Darstellen von Wissen*, ed. Wolfgang Krohn (Hamburg: F. Meiner, 2006).

87 Davis, "Profile of Benoît B. Mandelbrot," 7. Capitals in the original.

88 For a productive reflection on problems having to do with pictures in art and science, see Peter Geimer, *Bilder aus Versehen: eine Geschichte fotografischer Erscheinungen* (Hamburg: Philo Fine Arts, 2010), 294. Geimer quotes Gottfried Boehm's warning: "The maxims of a mere 'expansion'—of art, but of science as well—only lead to blurred simplifications"; Boehm, "Zwischen Auge und Hand," 52.

89 Mandelbrot, interview, April 2008.

90 See Gregory F. Lawler, Oded Schramm, and Wendelin Werner, "The Dimension of the Planar Brownian Frontier is 4/3," in Citebase, http://arxiv.org/abs/math/0010165v2.

91 Gregory F. Lawler, e-mail to the author, February 2, 2012.

92 Benoît Mandelbrot, "Fractals and the Geometry of Nature," in *1981 Yearbook of Science and the Future* (Chicago: Encyclopedia Britannica, Inc., 1980), 168.

93 Mandelbrot, *Les objets fractals*, 65. Mandelbrot compares the straight lines in the text with Baron Haussmann's redesign of Paris streets in the nineteenth century.

94 "En effet, il se propose d'établir qu'il existe entre tous ces objets de profondes analogies, se situant à un niveau intuitif qui mérite (malgré sa nouveauté) d'être considéré comme élémentaire."

95 For an introduction to picture arrangements as "picture tableaus," see Margarete Pratschke, "Bildanordnungen," in *Das Technische Bild. Kompendium zu einer Stilgeschichte wissenschaftlicher Bilder*, ed. Horst Bredekamp, Birgit Schneider, and Vera Dünkel (Berlin: Akademie, 2008), 116–19. For precursors of analogical thinking about pictures in the Middle Ages—again with a cosmological dimension—see Wolfgang Kemp, "Mittelalterliche Bildsysteme," *Marburger Jahrbuch für Kunstwissenschaft* 22 (1989). Analogy formation in the twentieth century has not been adequately studied. However, Barbara Stafford has presented an attempt; she accuses contemporary culture of impoverished analogical thinking insofar as it places greater emphasis on making distinctions than on recognizing similarities; Barbara Maria Stafford, *Visual Analogy: Consciousness as the Art of Connecting* (Cambridge, MA: MIT Press, 1999), 32.

96 Mandelbrot, interview, April 2008.

97 To the square of a number a one is added, the result squared, and again a one added, and so on, theoretically ad infinitum. Only numbers from this procedure that do not add up to an infinite value belong by definition to the set.

98 For a broader image-theoretical discussion of islands as mediums of separation and deviation and a draft of "islands as cultural technique," see Gloria Meynen, "Die Insel als Kulturtechnik (Ein Entwurf)," *Zeitschrift für Medienwissenschaft* 2 (2010): 79–91.

99 The term "polynomial" is a compound term for the type of equations under investigation here, with multiple variables and constants of a finite length. For the foundational works for this field, see Gaston Julia, "Mémoire sur l'itération des fonctions rationelles," *Journal de Math. pures et appliquées* 8.1 (1918): 47–245; and Pierre Fatou, "Sur les équations fonctionelles," *Bulletin de la Soc. Math. Fr.* 47 (1919): 161–271.

100 See Benoît Mandelbrot in A. Barcellos, "Interview of B. B. Mandelbrot," in *Mathematical People* (Boston: Birkhäuser, 1984), 211.

101 See John Hamal Hubbard, preface to *The Mandelbrot Set: Theme and Variations*, ed. Tan Lei (Cambridge: Cambridge University Press, 2000), xiii–xx.

102 Ibid., xv.

103 Benoît Mandelbrot, "Two Nonquadratic Rational Maps Devised from Weierstrass Doubling Formulas," in Mandelbrot, *Fractals and Chaos: The Mandelbrot Set and Beyond* (New York: Springer, 2004), 157.

104 Ibid.

105 Ibid.

106 Ibid., 161.

107 This phenomenon was described by the mathematician Samuel Lattès in 1918; see ibid., 139.

108 Ibid., 159.

109 Ibid., 169.

110 Ibid., 14 and 23.

111 For image interference and phantom images in a larger context, see Peter Geimer, "Was ist kein Bild? Zur 'Störung der Verweisung,'" in *Ordnungen der Sichtbarkeit. Fotografie in Wissenschaft, Kunst und Technologie*, ed. Peter Geimer (Frankfurt am Main: Suhrkamp, 2002), 313–41.

112 Mandelbrot, *Fractals and Chaos*, 14; and Mandelbrot, interview, April 2008.

113 It is interesting to note that Robert Brooks and Peter Matelski had presented their own picture of the Mandelbrot set at a mathematics conference the year before, though it was not published until three years later, together with the rest of the conference papers; see Brooks, Robert, and J. Peter Matelski, "The dynamics of 2-generator subgrounds of PSL (2, C)," in *Riemann*

Surfaces and Related Topics: Proceedings of the 1978 Stony Brook Conference, published in the *Annals of Mathematics Studies* 97, ed. Irwin Kra and Bernard Maski (Princeton: Princeton University Press, 1981), 71. In their ASCII version of the set, the bulbous shape is made up of individual little stars, with a lance-like extension on the left side. Brooks and Matelski were in this sense "discoverers without a discovery," for theirs was the first picture of such a set presented to the public, though they failed to recognize its self-similarity and complexity because the surrounding specks were too small to be seen with the visualization technology they used.

114 Mandelbrot, *Fractals and Chaos*, 14.

115 See Mandelbrot, *Les objets fractals*, 112–17.

116 See Nina Samuel, "'Do Not Clean Off the Dust Specks. They Are Real.' Über gestörte, verschmutzte und verborgene Computerbilder," in *Prekäre Bilder*, ed. Robert Suter and Thorsten Bothe (Munich: Wilhelm Fink, 2009), 19–47. I place Mandelbrot's proscriptions against cleaning in the context of various other ways in which computer pictures have been interfered with or become dysfunctional.

117 See Senechal, "The Continuing Silence of Bourbaki," 26.

118 Benoît Mandelbrot, "Fractal Aspects of the Iteration of $z \rightarrow \lambda z (1-z)$ for complex λ and z," *Nonlinear Dynamics: Annals of the New York Academy of Sciences* 357 (New York: New York Academy of Sciences, 1980), 250.

119 Mandelbrot, *Fractals and Chaos*, 22.

120 See ibid., 37.

121 Adrien Douady and John Hamal Hubbard, "Itération des polynômes quadratiques complexes," *Comptes rendus* (Paris) 294 (1982): 123: "Ces îlots sont en fait rattachés par des filaments qui échappent à l'ordinateur."

122 Adrien Douady, telephone interview by the author, January 7, 2005.

123 For a discussion of the "two planes" of the digital image see Frieder Nake, "Das doppelte Bild," *Bildwelten des Wissens, Kunsthistorisches Jahrbuch für Bildkritik* 3 (2005): 40–50.

124 Mandelbrot, *The Fractal Geometry of Nature*, text for color plate C1.

125 Thomas Schlich, "Repräsentationen von Krankheitserregern. Wie Robert Koch Bakterien als Krankheitsursache dargestellt hat," in eds. Hans-Jörg Rheinberger, Michael Hagner, and B. Wahrig-Schmidt, *Räume des Wissens. Repräsentation, Codierung, Spur* (Berlin: Akademie, 1997), 170.

126 See Benoît Mandelbrot, "Fractals and the Rebirth of Iteration Theory," in *The Beauty of Fractals: Images of Complex Dynamical Systems*, ed. Heinz-Otto Peitgen and Peter H. Richter (Berlin: Springer, 1986), 157; Mandelbrot, *Fractals and Chaos*, 19.

127 Ibid., 17.

128 Edward Hill, *The Language of Drawing* (Englewood Cliffs, NJ: Prentice-Hall, 1966), 25.

129 Douady, telephone interview. For the suppression of geometric drawings from official discourse in the mathematical sciences, see Peter Galison, "The Suppressed Drawing: Paul Dirac's Hidden Geometry," *Representations* 72 (2000): 145–66.

130 Douady, telephone interview.

131 The different variants of the "Hubbard tree" were discovered by John Hubbard in 1980; see Hubbard, preface to *The Mandelbrot Set: Theme and Variations*, xvi.

132 Plato, *Republic*, in *The Collected Dialogues of Plato*, ed. Edith Hamilton and Huntington Cairns, trans. Paul Shorey (New York: Pantheon, 1989), 744–47.

133 This is based on the interpretation by Wolfgang Kemp in "Disegno. Beiträge zur Geschichte des Begriffs zwischen 1547 und 1607," *Marburger Jahrbuch für Kunstwissenschaft* 22 (1989): 121–34.

134 Ibid., 230.

135 "Nuages de points tracés à l'ordinateur," in a letter to the author, December 20, 2004. John Hubbard had some forty variations of such "pictures of dot clouds."

136 Douady, telephone interview, and letter to author, December 20, 2004. The pictures showed "polynomial-like mappings" following a method that Douady had explained to Hubbard.

137 Ibid.

138 This name became a permanent term in mathematical literature. Douady also produced a didactic film on the "Dynamics of the Rabbit"; cf. Adrien Douady, Dan Sørensen, and François Tisseyre, *La dynamique du lapin* (Paris: Atelier Ecoutez voir, 1996).

139 Adrien Douady, "Systèmes dynamiques holomorphes," *Séminaire N. Bourbaki* 599 (1982): 58–60. For the Riemann sphere, which is thought of as a globe, see Hartmut Jürgens, Heinz-Otto Peitgen, and Dietmar Saupe, *Chaos: Bausteine der Ordnung* (Berlin: Springer, 1994), 361.

140 Douady, telephone interview.

141 For an art historical discussion of serpentines, see notes 49–51.

142 "Douady a dessiné les évolutions possibles des formes," Arnaud Chéritat, interview by author, Toulouse, February 8, 2008.

143 Dan Erik Krarup Sørensen, "Infinitely Renormalizable Quadratic Polynomials, with Non-Locally Connected Julia Sets," *The Journal of Geometric Analysis* 10 (2000).

144 Jean-Christophe Yoccoz (born 1957), French mathematician to whom the Fields Medal was awarded in 1994 for his work on dynamical systems.

145 "You have to study it like a tree," Douady, telephone interview by author, January 7, 2005.

146 Hilbert, "Mathematische Probleme," 27.

147 For another example of drawing as necessary for achieving mathematical understanding in this specific field, see Nina Samuel, "Beredte Skizzen. Chaos und das zerbrochene Ei von Yoshisuke Ueda," *Bildwelten des Wissens, Kunsthistorisches Jahrbuch für Bildkritik* 7, no. 2 (2010): 83–89.

148 Douady, "Systèmes dynamiques holomorphes," 62.

149 G. H. Hardy, "A Mathematician's Apology," in *The World of Mathematics*, Vol. 4, ed. James Roy Newman (New York: Simon and Schuster, 1956), 2027.

150 Joseph Beuys, *Zeichnungen 1947–59 I. Gespräch zwischen Joseph Beuys und Hagen Lieberknecht. Geschrieben von Joseph Beuys* (Cologne: Schirmer, 1972), 17. Beuys came up with many variations on this sentence over the course of his life.

151 Thomas Macho, "Das Rätsel der vierten Dimension," in *Science & Fiction. Über Gedankenexperimente in Wissenschaft, Philosophie und Literatur*, ed. Thomas

Macho and Annette Wunschel (Frankfurt am Main: Fischer, 2004), 77.

152 See Otto E. Rössler, "Chaos, Hyperchaos, and the Double-Perspective," in eds. Ralph Abraham and Yoshisuke Ueda, *The Chaos Avant-Garde: Memories of the Early Days of Chaos Theory* (Singapore: World Scientific, 2000), 209–19. For a discussion of his place in the history of chaos theory, see Christophe Letellier and Valério Messager, "Influences on Otto E. Rössler's Earliest Paper On Chaos," *International Journal of Bifurcation and Chaos* 20 (2010): 1–32.

153 For a historical overview with personal reports and important texts by "pioneers" in the field, see Abraham and Ueda, *The Chaos Avant-Garde*. For a further historiography see David Aubin and Amy Dahan Dalmedico, "Writing the History of Dynamical Systems and Chaos: Longue Durée and Revolution, Disciplines and Cultures," *Historia Mathematica* 29 (2002): 271–339.

154 For the first publication of the Rössler attractor see Otto E. Rössler, "Chaotic Behavior in Simple Reaction Systems," *Zeitschrift für Naturforschung* 31 (1976): 259–64.

155 See Edward Lorenz at MIT in 1963; see Edward N. Lorenz, "Deterministic Nonperiodic Flow," *Journal of the Atmospheric Sciences* 20 (1963): 130–41. Lorenz was a meteorologist and pioneer of chaos theory. He also coined the now-popular term "butterfly effect."

156 Otto E. Rössler, interview by author, Tübingen, Germany, June 1, 2007.

157 Ibid.

158 Ibid.

159 His goal was the transformation of Lorenz's model as a biochemical reaction system. The typical chemical-reaction diagrams with which he had been working for years served him as a starting point, and can be seen at the top of fig. 40.

160 Rössler thereby transferred one of the basic characteristics of chaotic dynamics onto his chemical clocks: folding and stretching topology; see Steve Smale, "Differentiable Dynamical Systems," *Bulletin of the American Mathematical Society* 73 (1967): 747–817.

161 In this film, produced a year later (1976), he can be seen operating the Dornier DO 240 analog computer that he used intensively from 1970 to 1979. See Letellier and Valérie Messager, "Influences," 4.

162 In order to arrive at a new understanding, his thinking had to be done "together with the folding and rolling of papers"; Rössler, interview.

163 Otto E. Rössler, "Chaos in Abstract Kinetics: Two Prototypes," *Bulletin of Mathematical Biology* 39 (1977): 282.

164 Bruce Stewart, interview by author, New York, May 14, 2008.

165 Rössler, interview.

166 Ibid.

167 Ibid.

168 "I cannot see what thinking should be fundamentally but just such tentative handling taking place only in the mind in some imaginary space"; Konrad Lorenz, *Der Abbau des Menschlichen. Die Rückseite des Spiegels. Versuch einer Naturgeschichte des Erkennens* (Munich: R. Piper, 1988), 175.

169 See Karin D. Knorr-Cetina, *Wissenskulturen. Ein Vergleich naturwissenschaftlicher Wissensformen* (Frankfurt am Main: Suhrkamp, 2002), 138.

170 Jacques Hadamard, *An Essay on the Psychology of Invention in the Mathematical Field* (New York: Dover, 1954), 76.

171 Rössler identified "all mental pictures as visual pictures as well," adding: "There are only visual pictures." Rössler, interview.

172 Ibid.

173 At the end of February 1882, Nietzsche wrote to Heinrich Köselitz (alias Peter Gast) on a typewriter that only had capital letters: "SIE HABEN RECHT—UNSER SCHREIBZEUG ARBEITET MIT AN UNSEREN GEDANKEN" ("You are right—our writing tools influence our thinking"; Friedrich Nietzsche, *Schreibmaschinentexte: Vollständige Edition, Faksimiles und kritischer Kommentar*, ed. Stephan Günzel and Rüdiger Schmidt-Grépály (Weimar: Bauhaus-Universität Weimar, Universitätsverlag, 2002), 18.

174 "In my room it is dead still—only the scratching of my pen on the paper—for I love to think while writing, since the machine has not yet been invented that registers our unexpressed, unwritten thoughts on any kind of material"; Friedrich Nietzsche, *Werke und Briefe. Historisch-kritische Gesamtausgabe*, vol. 2 of *Jugendschriften 1861–1864*, ed. Hans Joachim Mette (1933–40; Munich: Deutscher Taschenbuch, 1994), 71.

175 Rössler, interview.

176 Beuys, quoted from Magdalena Holzhey, *Im Labor des Zeichners: Joseph Beuys und die Naturwissenschaft* (Berlin: Reimer, 2009), 86.

177 Ibid., 87.

178 Merleau-Ponty, quoted from Georg Trogemann et al, "Experimente," in *Code und Material—Exkursionen ins Undingliche*, ed. Georg Trogemann (Vienna: Springer, 2010), 29.

179 Beuys, *Zeichnungen 1947–59,* 17.

Found Images I

Islands of Discovery: Sequentiality, Scribbles, and the Waste of Thought Processes

Nina Samuel

I automatically look for symmetries, identities, differences, which is an elementary way of handling pictures.[1]
—Benoît Mandelbrot

An important part of Benoît Mandelbrot's work revolves around the theme of islands, which exemplify in a twofold manner the use and the power of pictures in science. On the one hand, there is the attempt to generate figurative cartographies of chance and to stimulate the viewer's imagination (cats. 1–8); on the other, there are the experimental remains that preceded the discovery of the Mandelbrot set (cats. 9–39)—abstract patterns of insular detached dots or vague shadows. These scattered "island" structures, as Mandelbrot called them later with regard to his set, raise fundamental questions having to do with visualization, imaging technique, and mathematical realities: When is something visible mathematically existent, i.e., when can its existence be proven analytically? And when is its presence misleading? The items from Mandelbrot's office presented in the exhibition illustrate the productive, *poietic* (in the original Greek sense of "doing or making"), and seductive power of visual images as food for thought.

The pictures become meaningful as components of a series. Some even bear notations indicative of this. Either inscribed with the words "movie frame" or automatically stamped with the date on which they were produced—together with such additional technical details as the programmer's initials, the name of the algorithm represented, or its specific parameters—they are part of a timeline and clearly sequential. Essentially, a number of these visualizations were meant to show the solution a computer will find for a specific equation if multiple solutions are possible. The visualizations of mathematics access features that are not apparent from the formula alone. Each pixel corresponds to a complex number, and hence, on the monitor, a precisely defined surface point on the complex plane. The coloring of the pixels marks the different behavior of numerical values according to a previously established number of calculation steps. With these patterns, Mandelbrot saw himself confronted by an overwhelming variety. They were difficult to interpret, and this affected the progress of the experiments. Frequently, a sense of "increasing confusion" was what "motivated the next step."[2] As part of a series of more than 150 graphics whose theoretical basis was as yet not understood, they appeared at first glance to be scientific "detritus," though in reality they were very important, revealing intermediate steps along the way to the discovery of the Mandelbrot set and documenting the challenge of dealing with a type of imagery new to the history of mathematics: pictures that had to be observed.

As the dominant pictorial form, the sequence made it possible to compare patterns as a means of attaining mathematical understanding. Often, the blankness of a page was significant and highly revealing. The shapes that appeared, changed, and disappeared again were not only considered in themselves, but also in relation to neighboring images. It was only thanks to the previous and subsequent images that they made sense; without them, they would have been incomprehensible. A fundamental feature of scientific images is that they constitute both detailed individual shapes to be considered separately and testaments to a process; they are shapes that refer to pictorial traditions and traces that are always part of a chain of representation.[3]

Among the finds from Mandelbrot's office are various scribbles that document his preoccupation with the composition of his islands (cats. 40–48) and sketches that he produced for his programmers. Yet he always made it a point to stress that drawings made by hand were only of subordinate interest for him: "Drawing was not important. . . . The drawing period ended very early."[4] Instead, he would praise the precision of the computer: "Hand drawing would have been prohibitive, but I was fortunate in having access to an experimental computer-graphics device that produces camera-ready copy."[5] While Mandelbrot publically marginalized drawing as a method for gaining mathematical understanding, there was a second type of imaging whose importance he emphasized: "I draw in my head."[6] His co-workers confirm how important these "mind drawings" were to his work—the programmer Alan Norton, for example, said "I think that most of those sketches existed in his mind, and that is often the way it is with mathematicians. The scribbles are throw-away sketches, discarded after the work is complete."[7] Some of these supposed "by-products," these "throw-away sketches," could nonetheless be salvaged for the exhibition. They document the feedback-loops created between writing utensil, paper, and thought; the clumsy eloquence of the material; and the way reasoning took place on paper.

I would like to thank Andrew Goodhouse for his thoughtful contributions to our conversations about repetition and variation.

1 Benoît Mandelbrot, see interview in this volume, p. 156.
2 Benoît Mandelbrot, *Fractals and Chaos: The Mandelbrot Set and Beyond* (New York: Springer, 2004), 161.
3 See Bruno Latour, "Circulating Reference: Sampling the Soil in the Amazon Forest," in *Pandora's Hope: Essays on the Reality of Science Studies* (Cambridge, MA: MIT Press, 1999). For the fundamental dichotomy in scientific pictures—the way they are always part of a formal tradition yet also always establishing differences—see Jochen Hennig, "Bildtradition und Differenz. Visuelle Erkenntnisgewinnung in der Wissenschaft am Beispiel der Rastertunnelmikroskopie," in *Das Technische Bild. Kompendium zu einer Stilgeschichte wissenschaftlicher Bilder* (Berlin: Akademie Verklag, 2008).
4 Benoît Mandelbrot, telephone interview by author, October 12, 2004.
5 Benoît Mandelbrot, *Fractals: Form, Chance, and Dimension* (San Francisco: Freeman, 1977), 24.
6 Benoît Mandelbrot, interview by author, Cambridge, MA, April 2008.
7 Alan Norton, e-mail to the author, January 29, 2012.

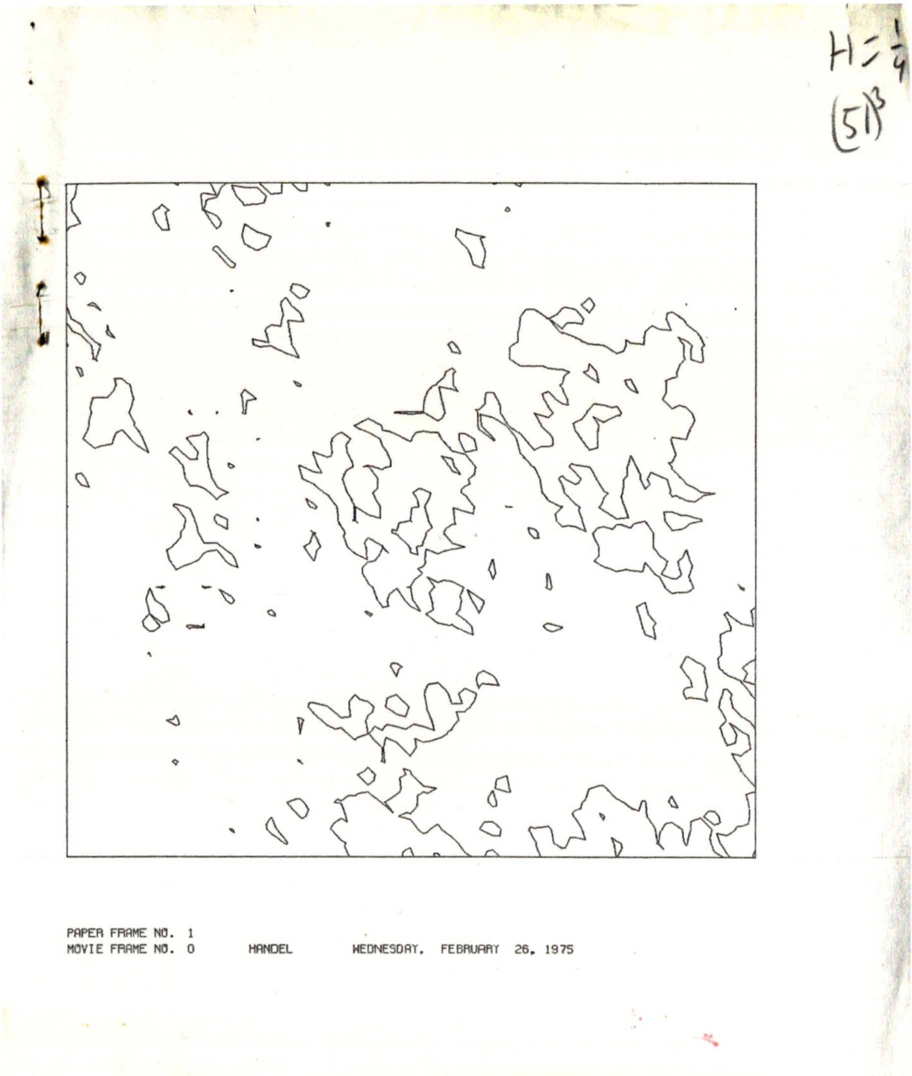

Cats. 1–4. Benoît Mandelbrot and Sigmund Handelman, programmer. "Islands," 1975. Computer-generated prints. Collection Aliette Mandelbrot.

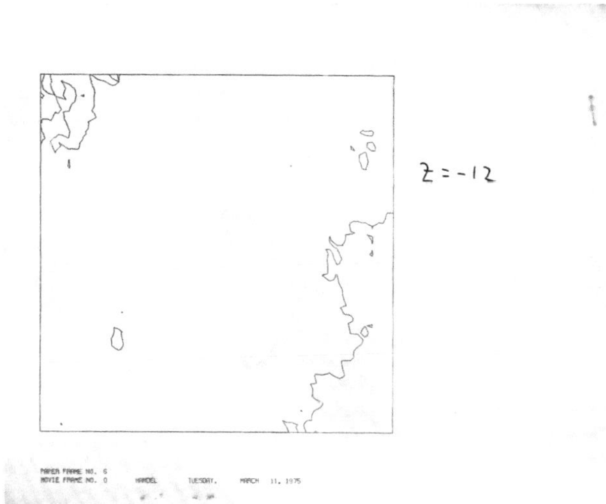
Z = -12
PAPER FRAME NO. 6
MOVIE FRAME NO. 0 HANDEL TUESDAY. MARCH 11, 1975

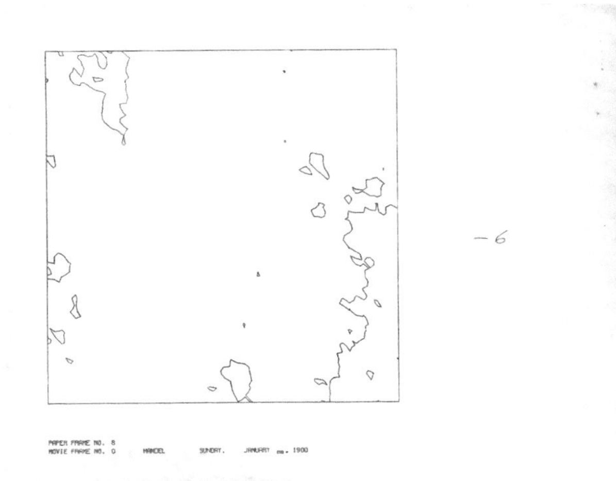
-6
PAPER FRAME NO. 8
MOVIE FRAME NO. 0 HANDEL SUNDAY. JANUARY 1900

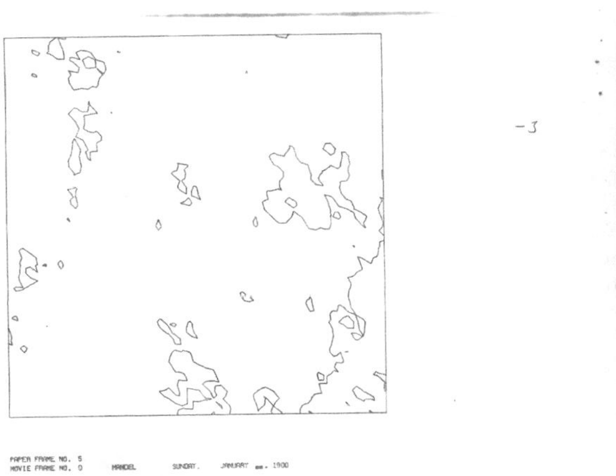
-3
PAPER FRAME NO. 5
MOVIE FRAME NO. 0 HANDEL SUNDAY. JANUARY 1900

S. HANDELMAN Z= -1.0

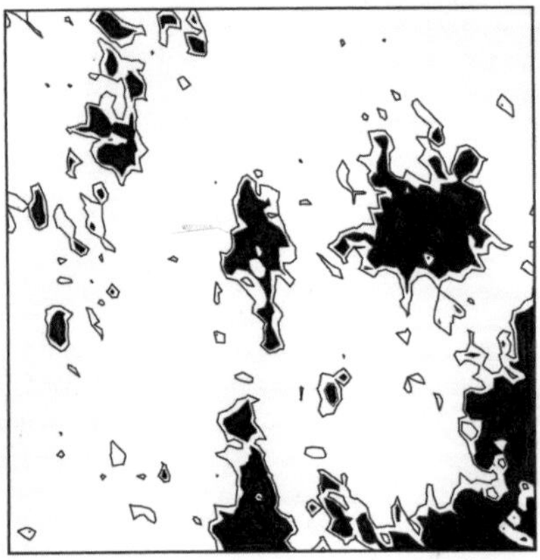

S. HANDELMAN Z= -5.0

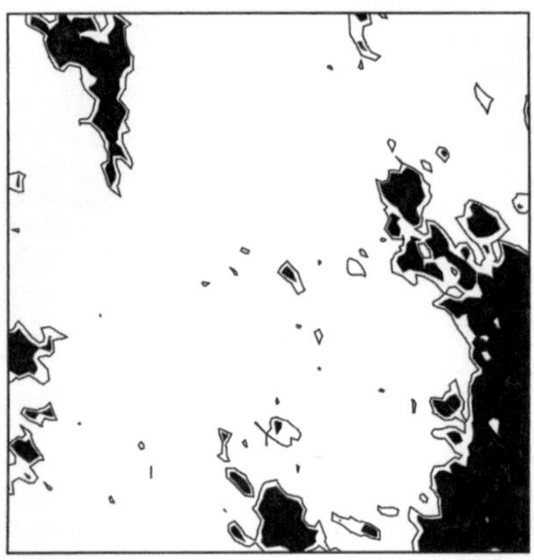

S. HANDELMAN Z=-15.0

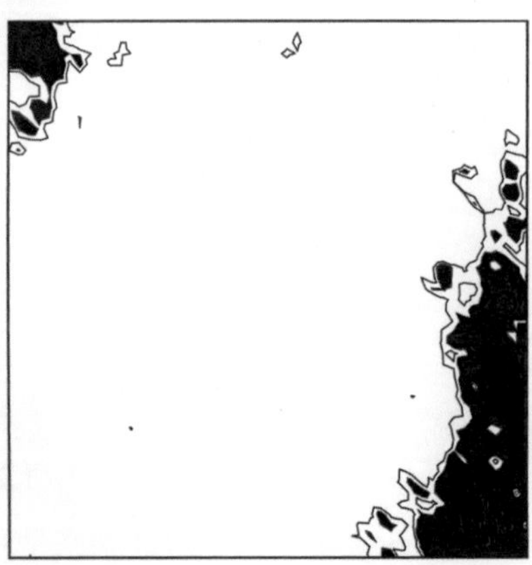

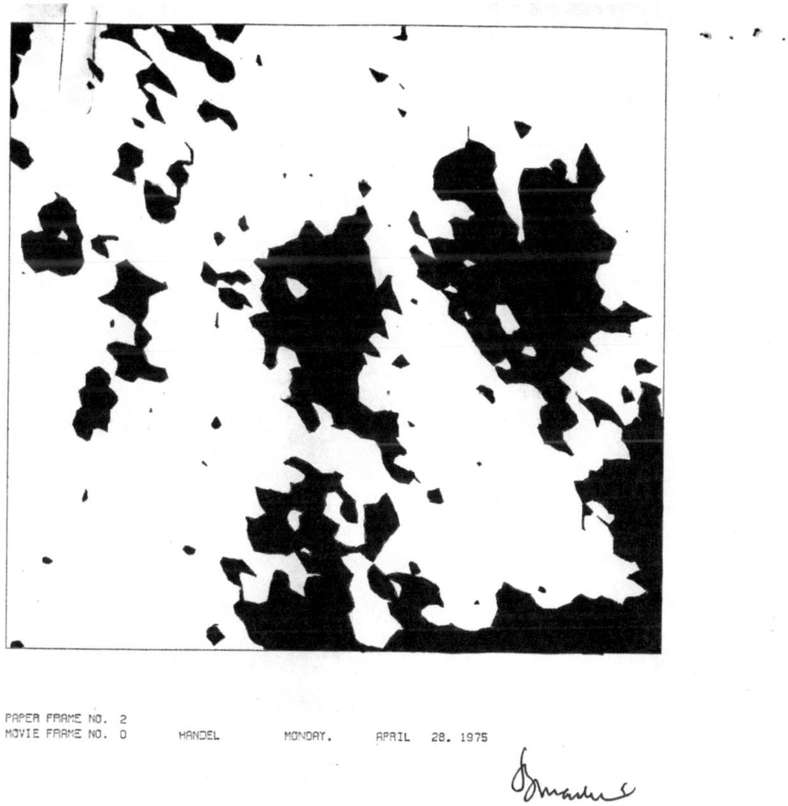

Cats. 5–8. Benoît Mandelbrot and Sigmund Handelman, programmer. “Islands,” 1975. Hand-colored computer-generated prints. Collection Aliette Mandelbrot.

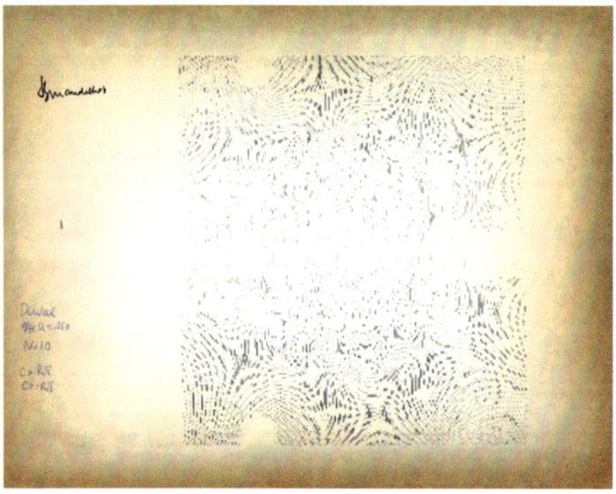

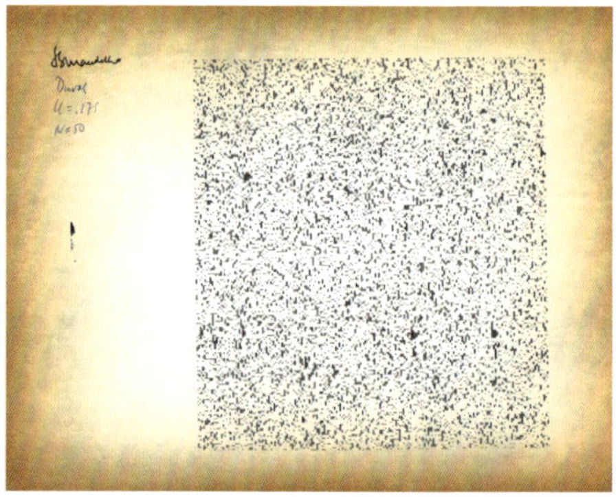

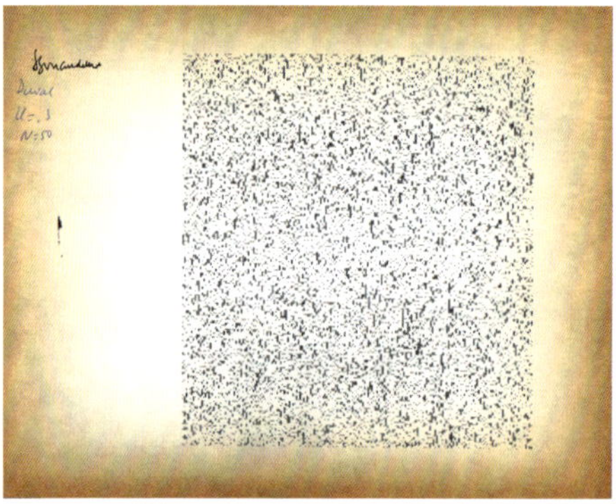

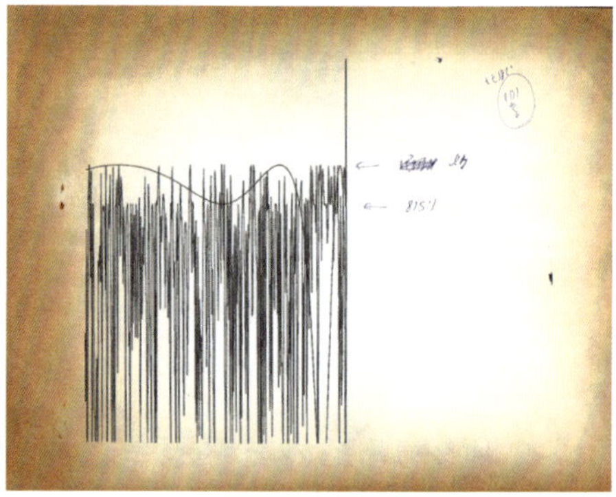

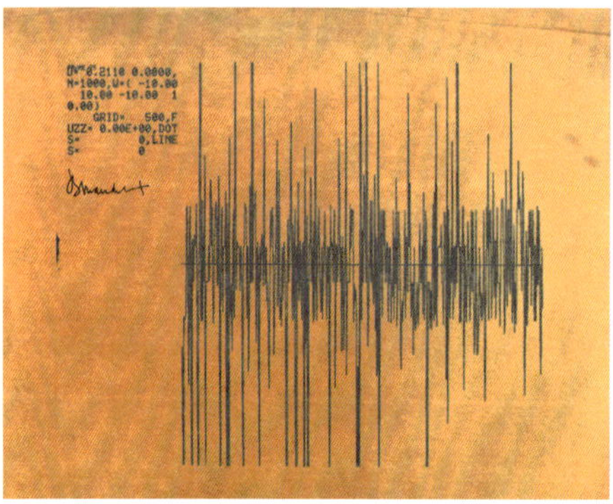

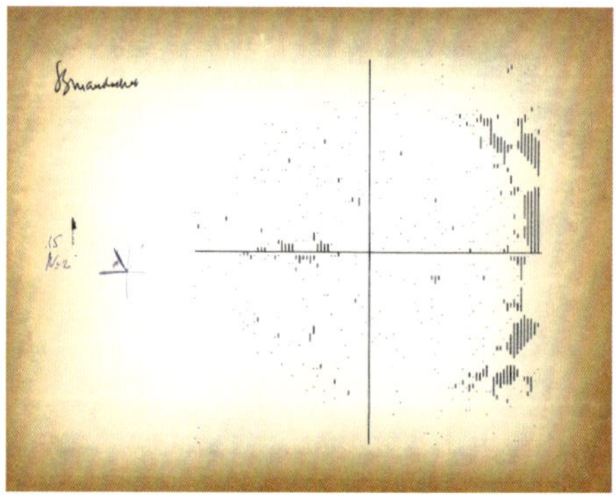

Cats. 9–17. Benoît Mandelbrot. Investigations in the complex plane, ca. 1979–80. Poss. Harvard or IBM. Computer-generated thermographic prints. Collection Aliette Mandelbrot.

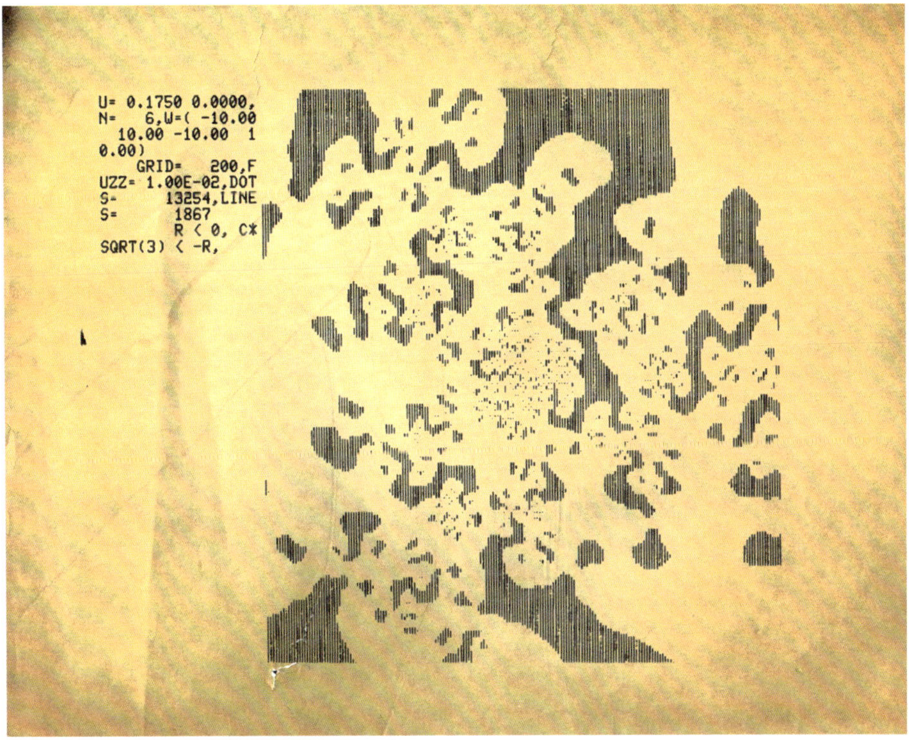
U= 0.1750 0.0000,
N= 6,W=(-10.00
10.00 -10.00 1
0.00)
GRID= 200,F
UZZ= 1.00E-02,DOT
S= 13254,LINE
S= 1867
R < 0, C*
SQRT(3) < -R,

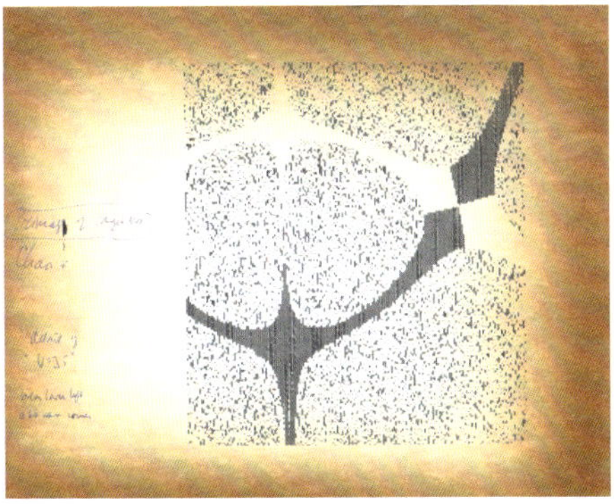

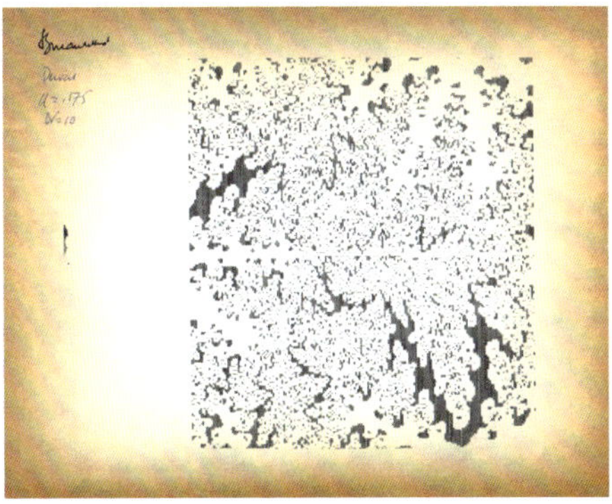

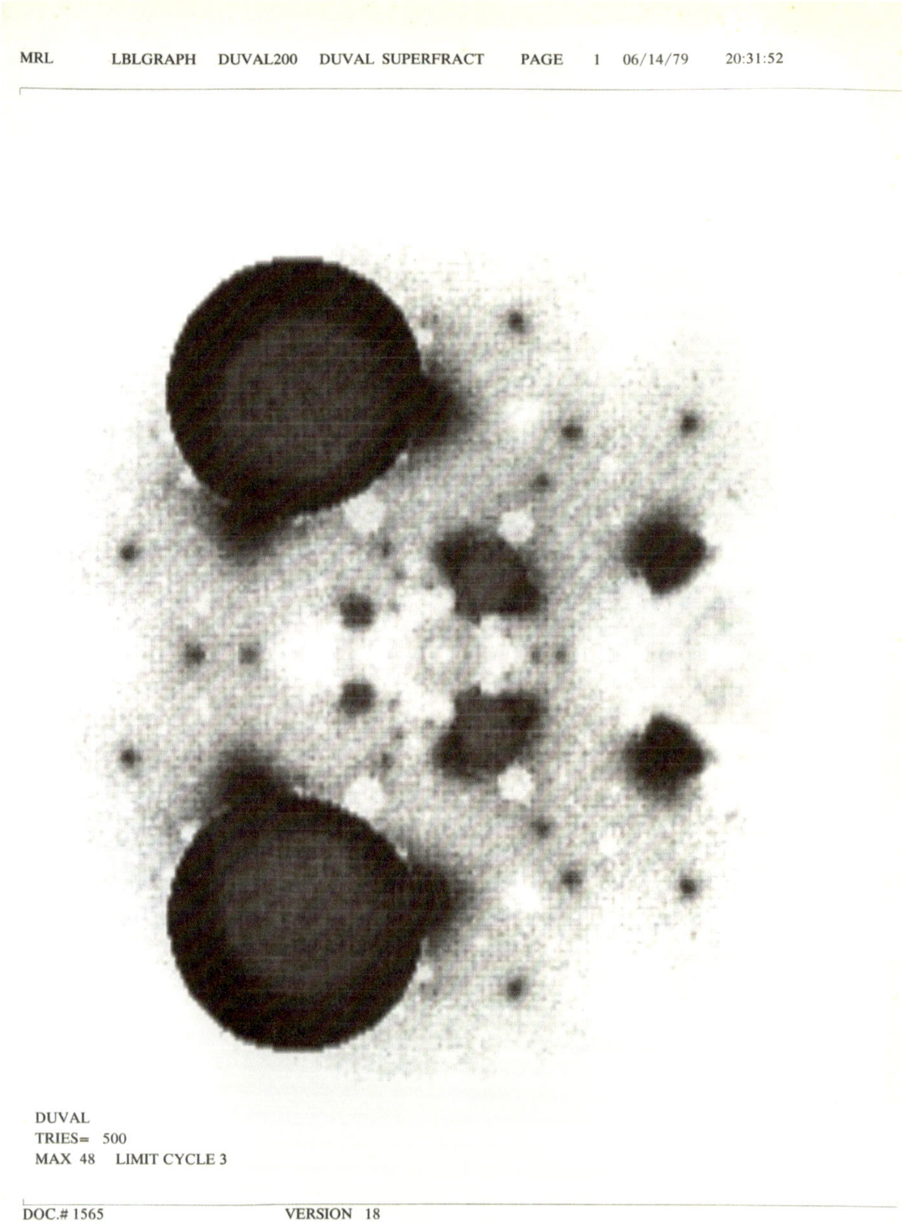

Cats. 18–21. Benoît Mandelbrot and Mark R. Laff, programmer. Investigations in the complex plane of iterated polynomials, series of over 150, 1979. IBM. Computer-generated prints on photographic paper. Collection Aliette Mandelbrot.

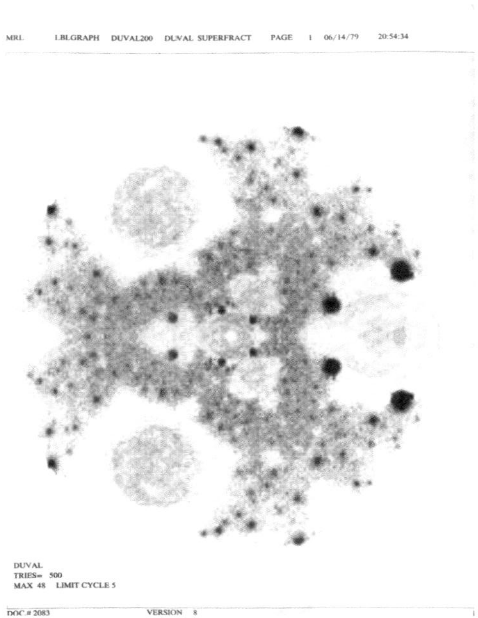
MRL LBLGRAPH DUVAL200 DUVAL SUPERFRACT PAGE 1 06/14/79 20:54:34
DUVAL
TRIES= 500
MAX 48 LIMIT CYCLE 5
DOC.# 2083 VERSION 8

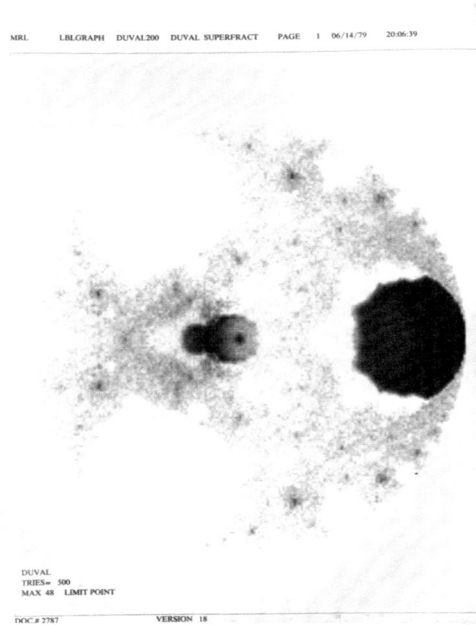
MRL LBLGRAPH DUVAL200 DUVAL SUPERFRACT PAGE 1 06/14/79 20:06:39
DUVAL
TRIES= 500
MAX 48 LIMIT POINT
DOC.# 2787 VERSION 18

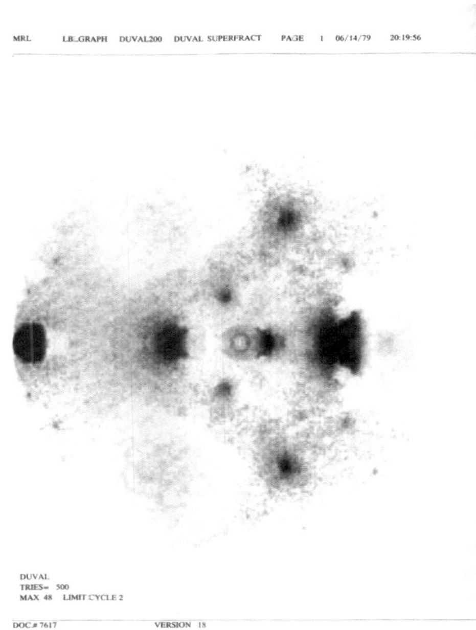
MRL LBLGRAPH DUVAL200 DUVAL SUPERFRACT PAGE 1 06/14/79 20:19:56
DUVAL
TRIES= 500
MAX 48 LIMIT CYCLE 2
DOC.# 7617 VERSION 18

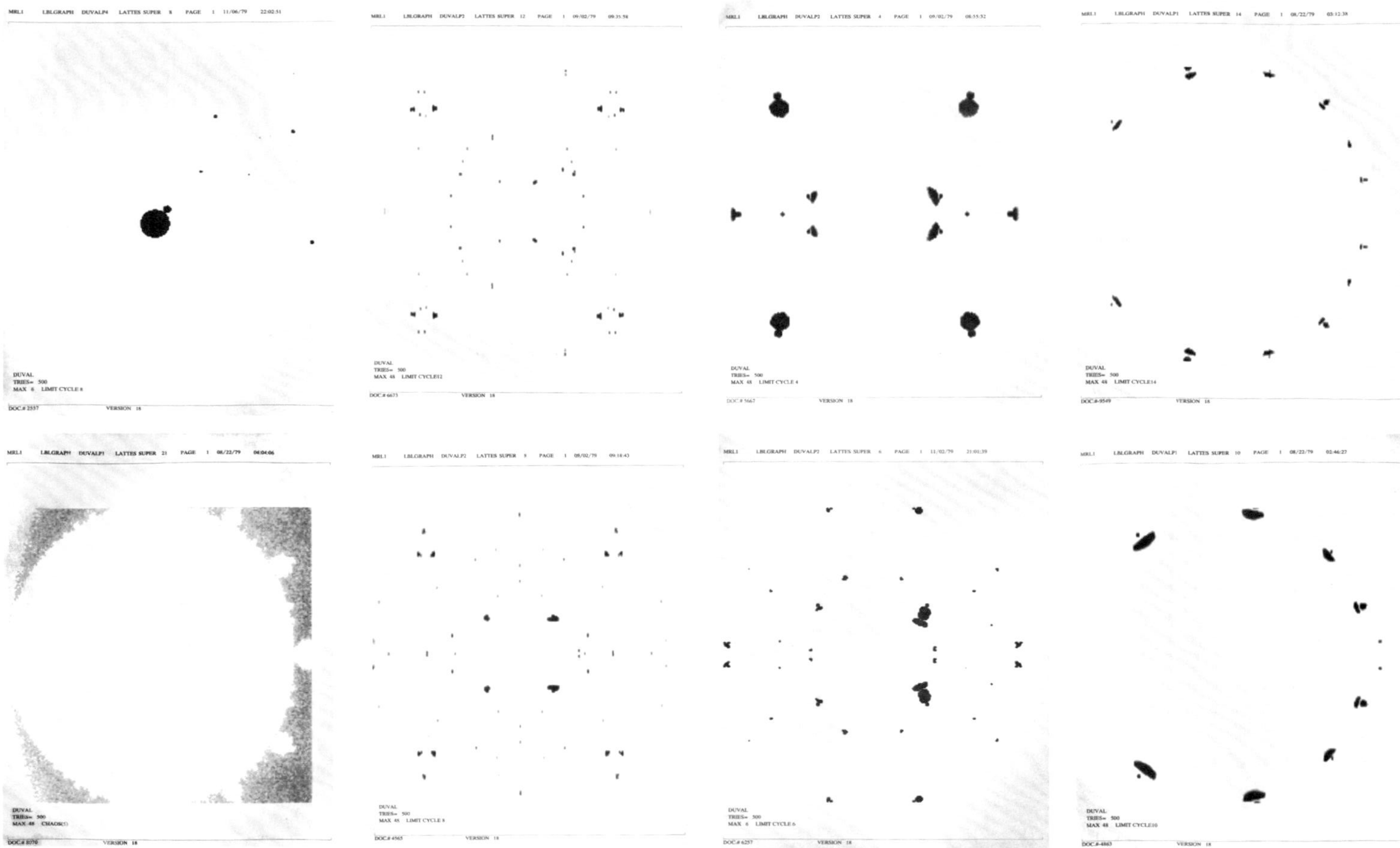

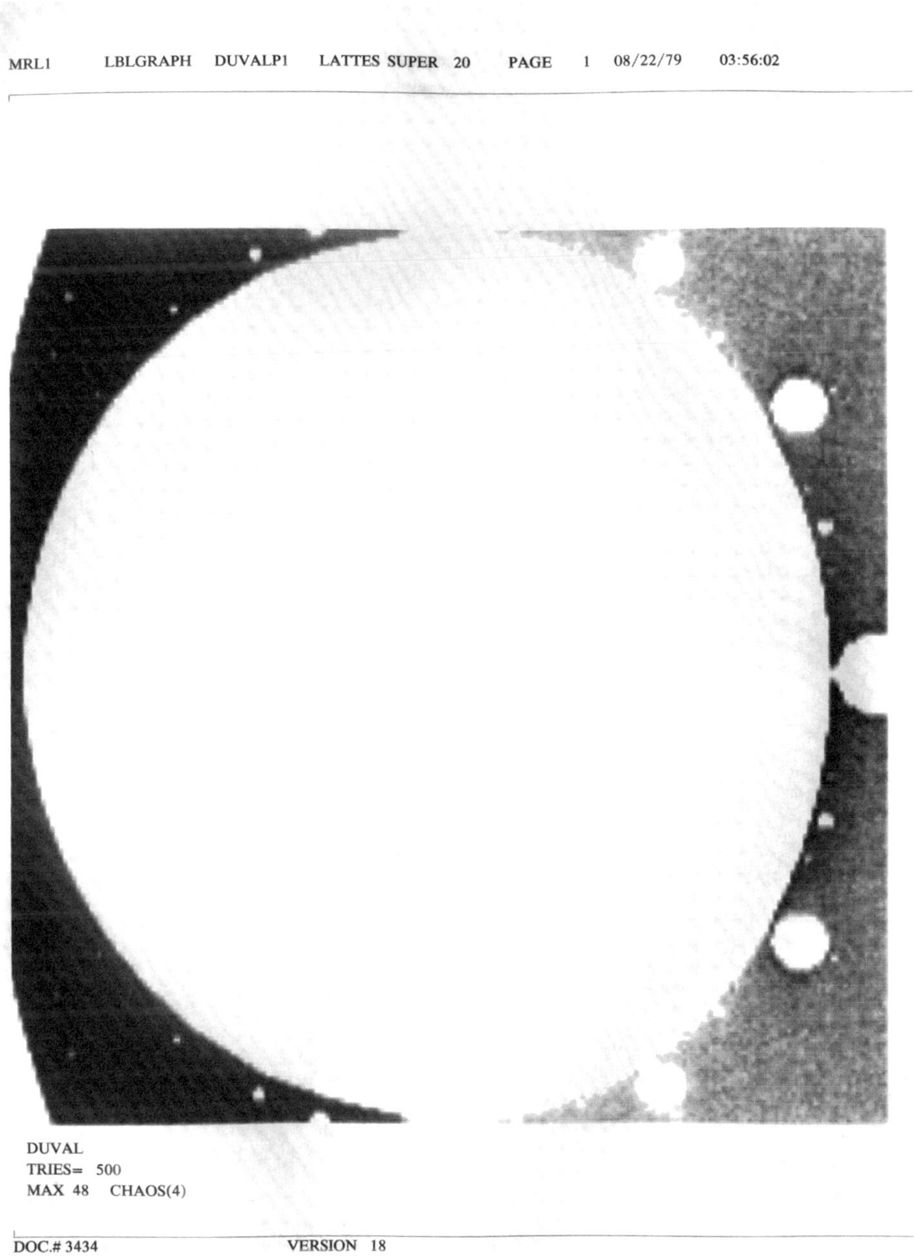

Cats. 22–30. Benoît Mandelbrot and Mark R. Laff, programmer. Investigations in the complex plane of iterated polynomials, series of over 150, 1979. IBM. Computer-generated prints on photographic paper. Collection Aliette Mandelbrot.

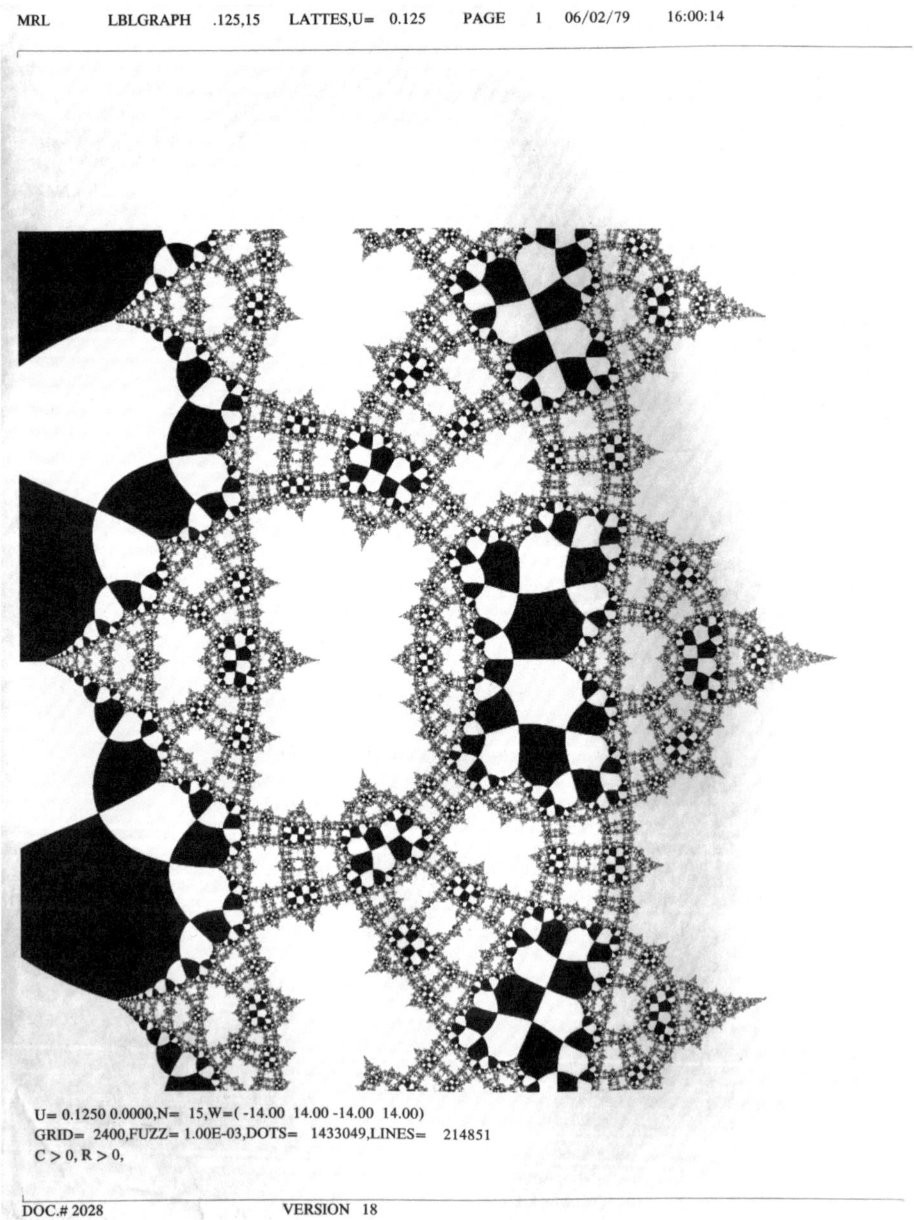

Cats. 31–39. Benoît Mandelbrot and Mark R. Laff, programmer. Investigations in the complex plane of iterated polynomials, series of over 150, 1979. IBM. Computer-generated prints on photographic paper. Collection Aliette Mandelbrot.

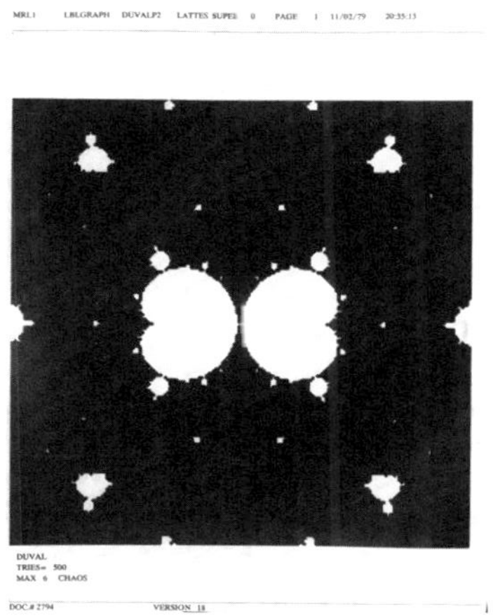
MRL1 LBLGRAPH DUVALP2 LATTES SUPER 0 PAGE 1 11/02/79 20:35:13
DUVAL
TRIES= 500
MAX 6 CHAOS
DOC.# 2794 VERSION 18

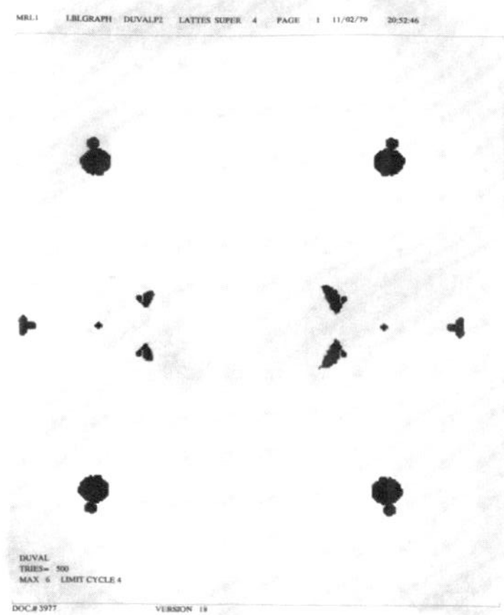
MRL1 LBLGRAPH DUVALP2 LATTES SUPER 4 PAGE 1 11/02/79 20:52:46
DUVAL
TRIES= 500
MAX 6 LIMIT CYCLE 4
DOC.# 2977 VERSION 18

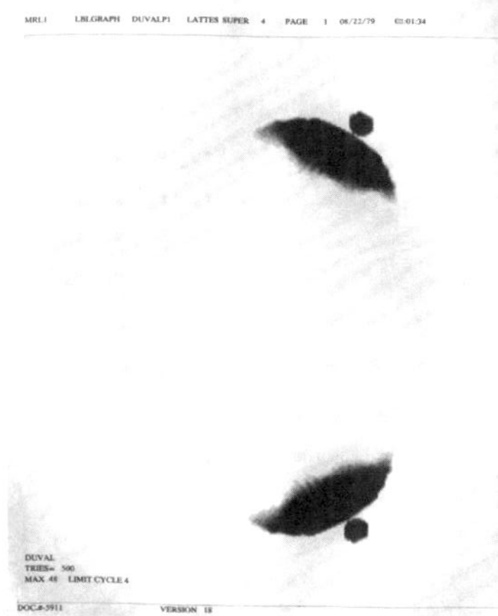
MRL1 LBLGRAPH DUVALP1 LATTES SUPER 4 PAGE 1 08/22/79 02:01:34
DUVAL
TRIES= 500
MAX 48 LIMIT CYCLE 4
VERSION 18

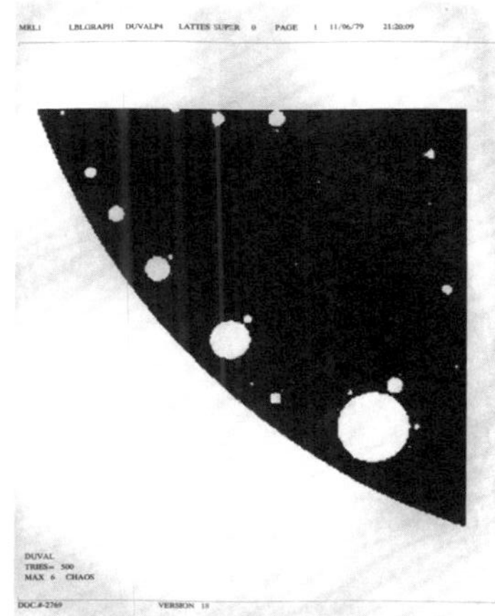
MRL1 LBLGRAPH DUVALP4 LATTES SUPER 0 PAGE 1 11/06/79 21:20:09
DUVAL
TRIES= 500
MAX 6 CHAOS
VERSION 18

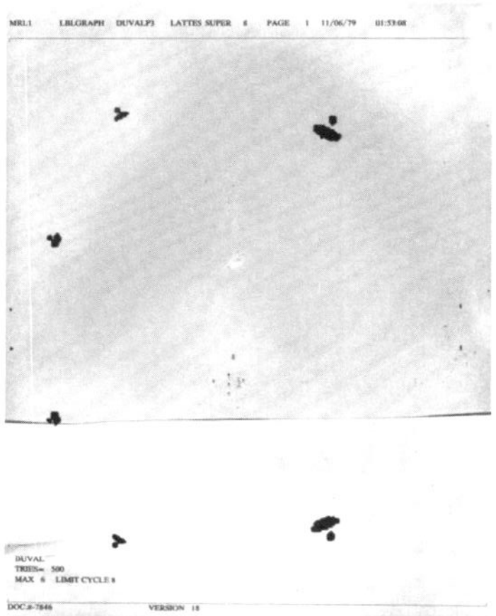
MRL1 LBLGRAPH DUVALP3 LATTES SUPER 8 PAGE 1 11/06/79 01:53:08
DUVAL
TRIES= 500
MAX 6 LIMIT CYCLE 8
VERSION 18

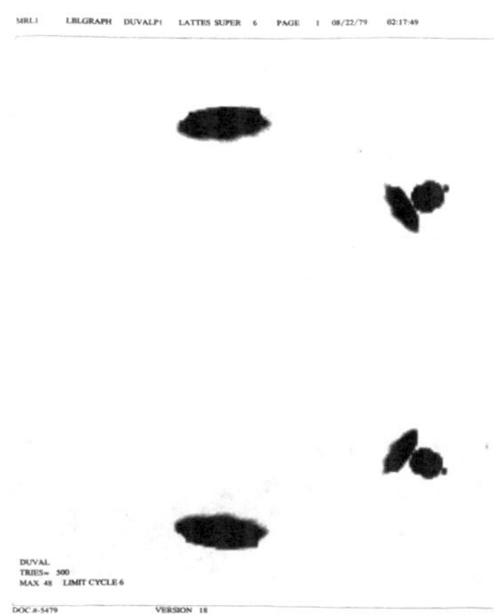
MRL1 LBLGRAPH DUVALP1 LATTES SUPER 6 PAGE 1 08/22/79 02:17:49
DUVAL
TRIES= 500
MAX 48 LIMIT CYCLE 6
DOC.#-5479 VERSION 18

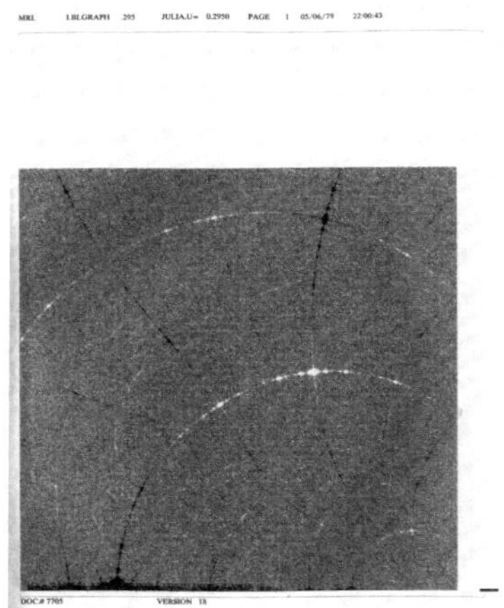
MRL LBLGRAPH 295 JULIA,U= 0.2950 PAGE 1 05/06/79 22:00:43
DOC.# 7705 VERSION 18

MRL1 LBLGRAPH DUVALP3 LATTES SUPER 1 PAGE 1 09/05/79 21:54:42
DUVAL
TRIES= 500
MAX 48 LIMIT POINT
VERSION 18

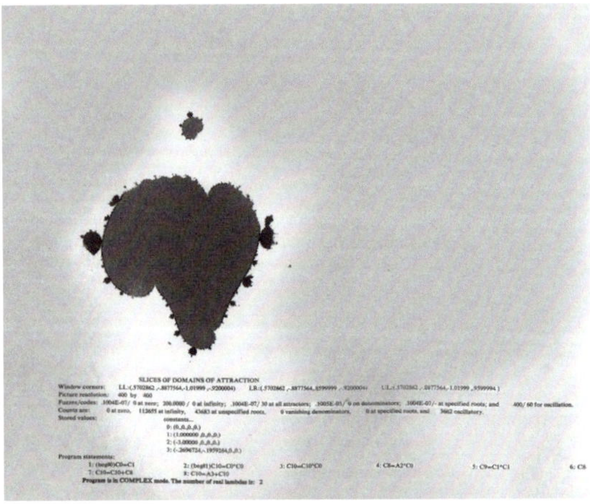
SLICES OF DOMAINS OF ATTRACTION
Program is in COMPLEX mode. The number of real lambdas is: 2

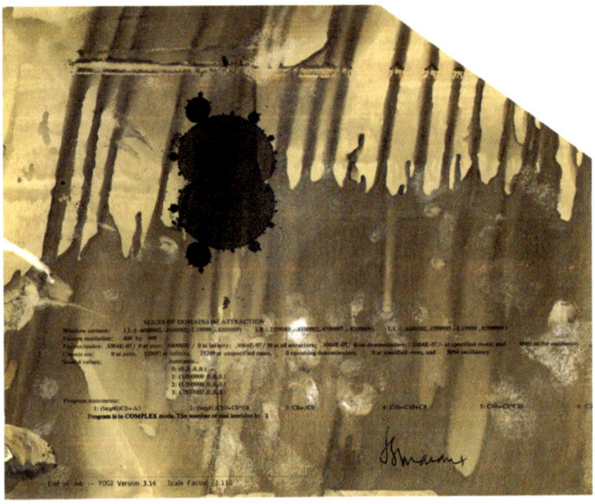
End of Job -- YOGI Version 3.14 Scale Factor 1.110

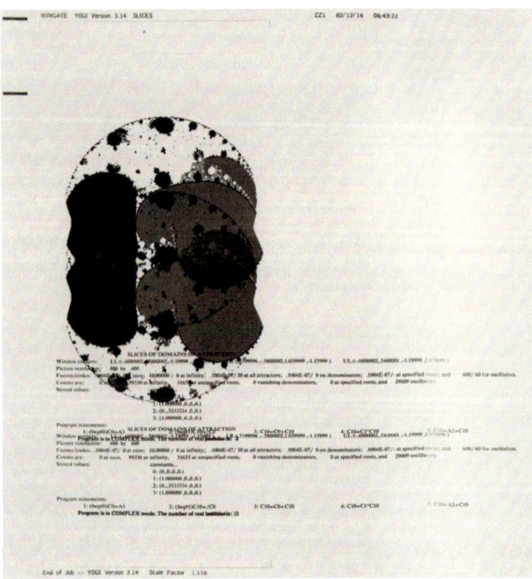
WINGATE YOGI Version 3.14 SLICES
ZZ1 82/12/16 06:43:22
End of Job -- YOGI Version 3.14 Scale Factor 1.110

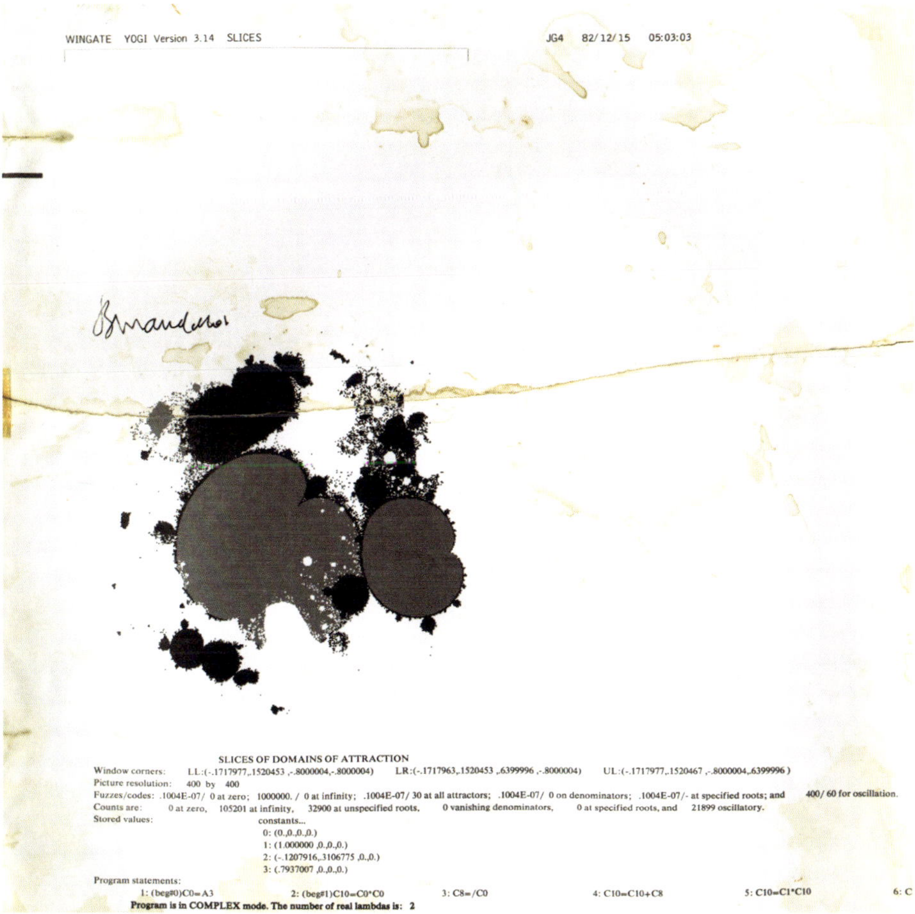

Cats. 40–43. Benoît Mandelbrot and Alan Norton, programmer. "Erroneous" and "destroyed" variations on the Mandelbrot set, ca. 1982–83. Computer-generated prints and misprints on photographic paper, torn and stained. Collection Aliette Mandelbrot.

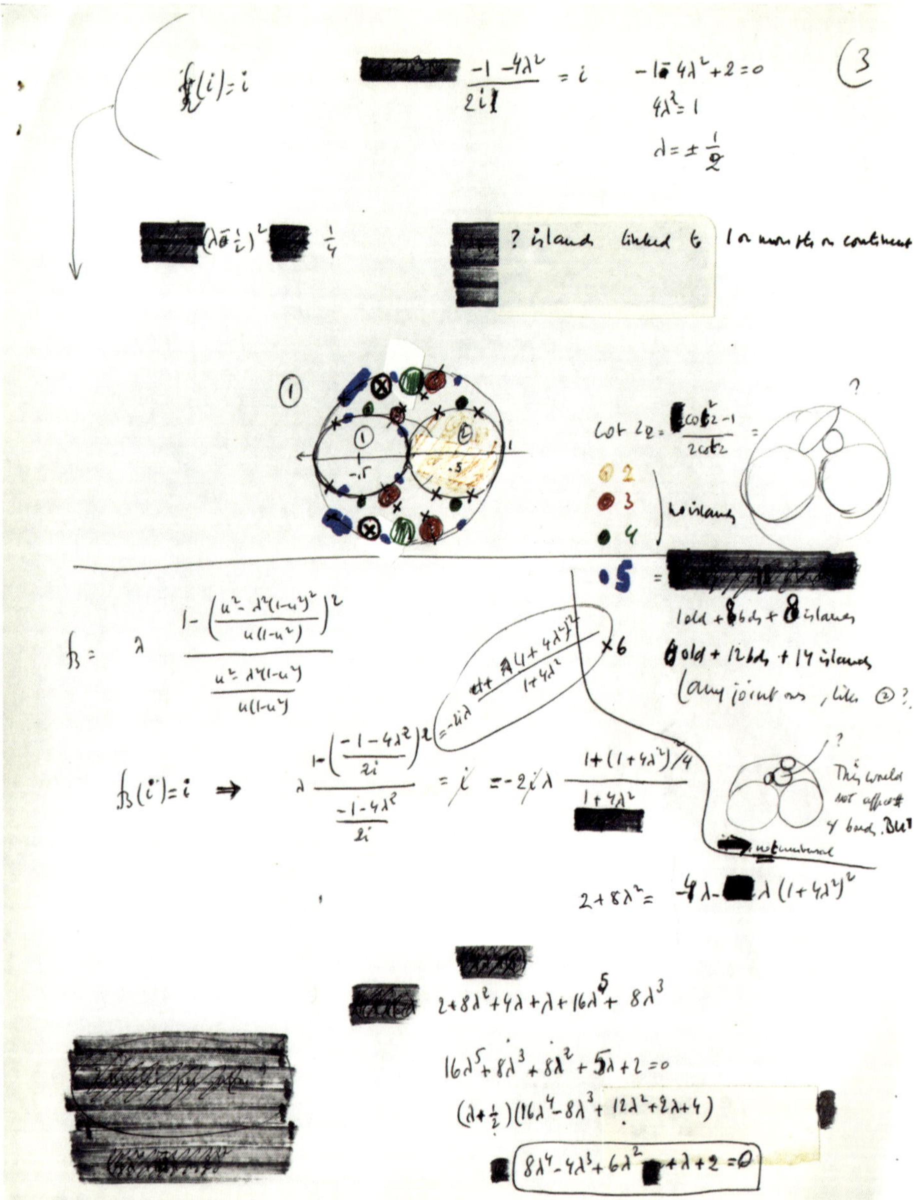

Cat. 44. Benoît Mandelbrot. Scribbles on the topic of the Mandelbrot set and related to the question of the connectedness of the "islands," after 1980. Pen and marker on paper. Collection Aliette Mandelbrot.

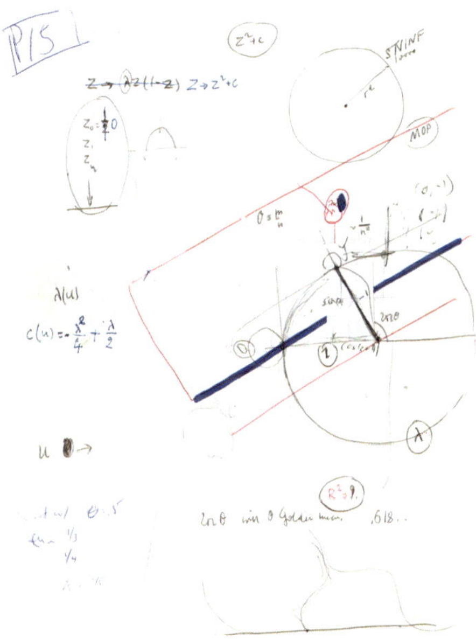

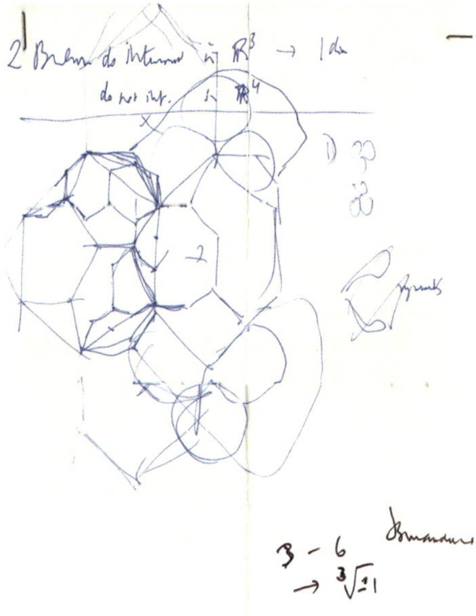

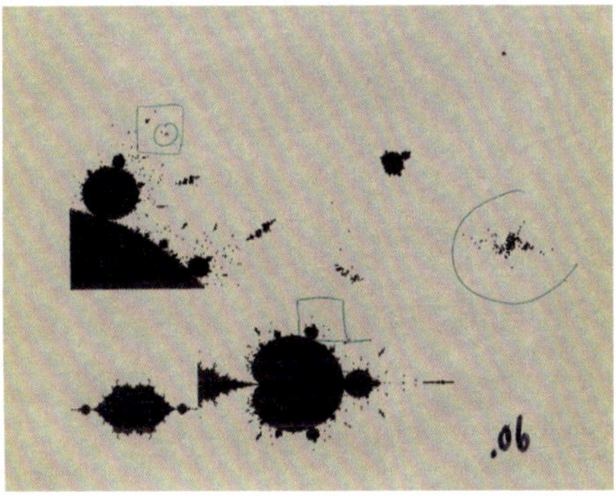

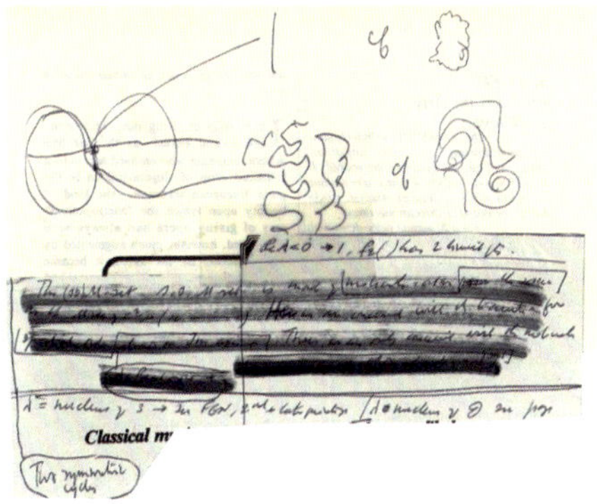

Cat. 45 (top left) Benoît Mandelbrot. Scribbles on the topic of the Mandelbrot set, after 1980. Pen and marker on paper. Collection Aliette Mandelbrot.

Cat. 46 (top right) Benoît Mandelbrot. Scribbles, after 1976. Pen on cardboard. Collection Aliette Mandelbrot.

Cat. 47 (bottom left) Benoît Mandelbrot. Investigation of the speckled "islands" of the Mandelbrot set, after 1980. Computer-generated print with scribbles. Collection Aliette Mandelbrot.

Cat. 48. (bottom right) Benoît Mandelbrot. Scribbles, undated. Pen and marker on paper. Collection Aliette Mandelbrot.

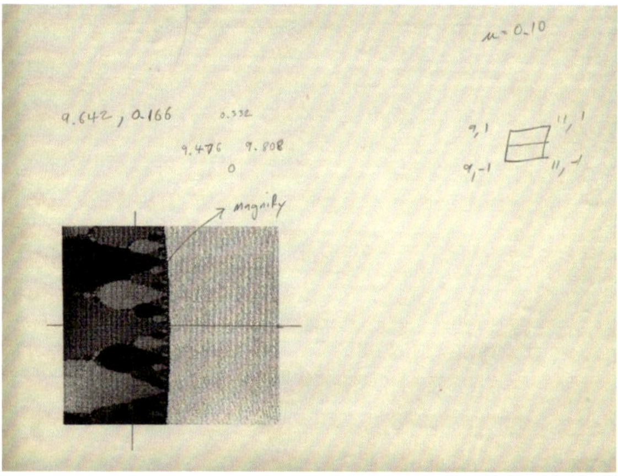

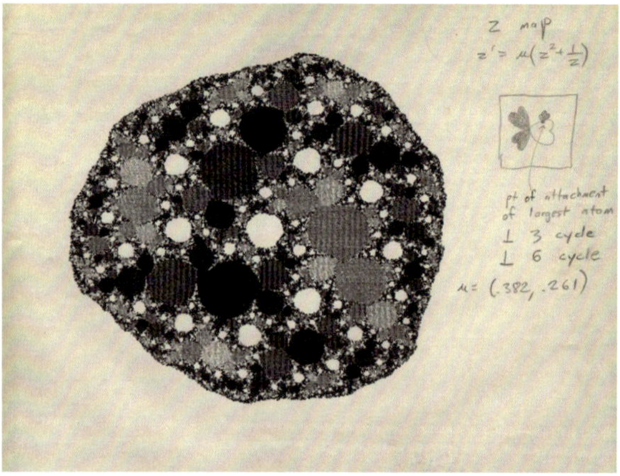

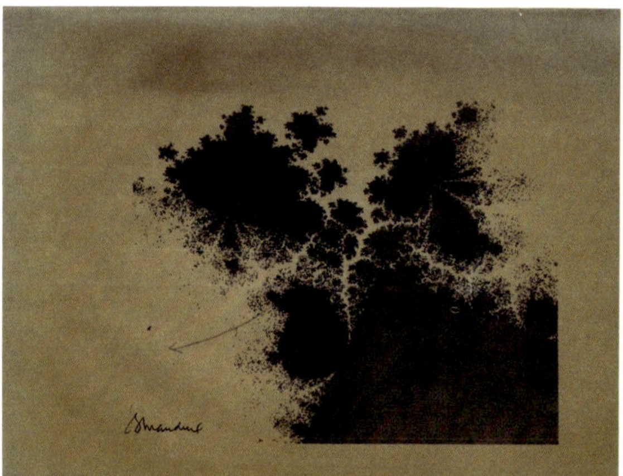

Cats. 49–51. Benoît Mandelbrot. Scribbled arrows (probably from a co-worker) showing the method by which details are magnified. Computer-generated prints with scribbles, undated. Collection Aliette Mandelbrot.

Pictures of a Higher Order

Matthias Bruhn

Fighting the Chaos

Chaos comes unbidden, but banishing it takes work. If one can no longer find anything on one's writing table, if freedom threatens to tip over into anarchy, or if processes come to a halt, the term "chaos" is readily at hand; one speaks of "traffic chaos," for example, or "chaotic situations," and with it there is generally the suggestion that someone with vision and competence will swiftly set about the job of clearing it up.

Chaos and fighting chaos are accordingly key concepts in politics. They legitimize political decisions and interventions and they give rise to political imagery: on one side, squalor and confusion, angry demonstrators, and burning streets; on the other side, strong hands and determined men of steel, intact families, and airy architecture. Opposed to the need for clarity and hygiene is the fear of darkness and infection, which is also transferred in propaganda and consumer advertising onto lifestyles, ethnic groups, and concepts of the enemy. In an attempt to make chaos a battle cry of politics, a wide-ranging and contradictory iconography of power has been developed. Without it, modern "chaos theory" might not have attracted the kind of media attention it was met with in the beginning, for anyone who masters chaos possesses power.

In political propaganda, for example, one frequently encounters motifs of order through cleaning, as in the old saw, "A new broom sweeps clean."[1] The theme appears in many variations throughout the nineteenth and twentieth centuries, as in Victor Deni's famous poster "Comrade Lenin Cleans the Earth of Scum" from 1920, in which the righteous Communist leader sweeps feudal-era tycoons and monarchs off the surface of the earth (fig. 1). In a work of Deni's from 1943, a Red Army soldier dispatches German occupation forces using a broom of bayonets. A few years after the demise of

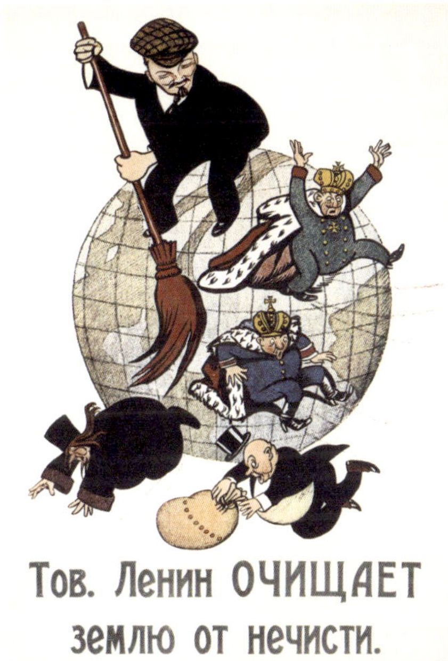

Fig. 1. V. N. Deni. "Comrade Lenin Purging the Unclean from the Earth," ca.1920. Color silkscreen, original strike. Permanent Collection of the Nassau County Museum of Art, Gift of the Saltzman Family Foundation.

the Nazi regime, which had prophesied that it would "triumph over Bolshevist Chaos," a West German party was already campaigning for "cleanliness in the state" using a simplified drawing of a broom symbolizing the desired anti-Communist cleanup work.[2]

Images like these imply that only he who employs a degree of firmness and ruthlessness is capable of changing things. But it is not only in Deni's posters that such a figure is portrayed as a political giant, one to whom a silent or noisy majority submits and is seen to act on the majority's behalf, creating a new society, a new order.

World Order, Visual Order

For millennia, the dialectic of order and chaos has repeatedly brought forth new pictorial forms designed to find a logical place for every part of creation. To ward off the wild and untamed, for example, monsters and mutants began to appear on the capitals and portals of medieval Christian churches and convents. Visual models were long used to create synoptic overviews and systematic contexts, as in diagrams of the cosmos (which schematized man's shared knowledge), family trees (which documented man's divine origin), and world maps (which took a wide shot of the world and banished chaos to the margins as *terra incognita* or uncharted ocean).

There are other important components of order. In Christian depictions of the Last Judgment and earthly justice, for example, regularity, uniformity, and symmetry stand for clear, hierarchical relationships. Heavenly and earthly judgment are directly juxtaposed in a double picture from 1478 in the Stadtmuseum Graz, where the town magistrate, Niklas Strobel, sits next to a panel of heavenly judges (fig. 2). The work's horizontal symmetry serves to guarantee an appropriate verdict, its vertical symmetry an equitable one. Every element is presented in a specific sequence and size, and the figures are at a uniform scale.

Similar principles of balance and rhythm continue to govern the exterior and interior architecture of parliaments, courts of justice, convents, and temple districts throughout the world. They are principles that are intended to evoke equity and stability, to provide orientation, and to point to a higher authority. In their severe regularity, they preclude any sort of uncertainty.

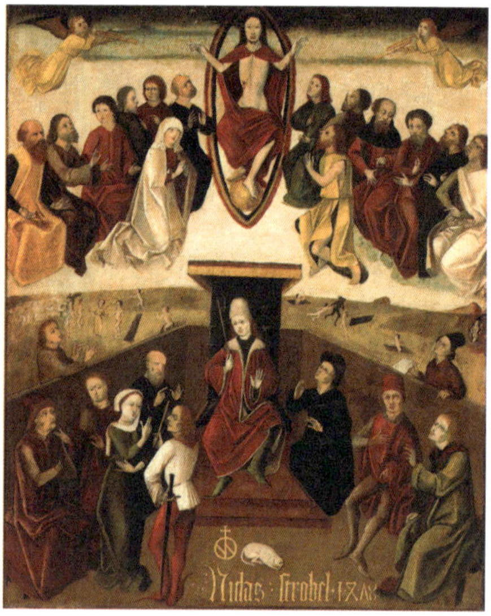

Fig. 2. Anonymous. *Gerechtigkeitsbild* (Painting of justice), 1478. Oil, tempera on spruce. Stadtmuseum graz GmbH.

In Western visual culture, central perspective is another means by which authority can be symbolized; with it, the visible world is oriented toward a vanishing point and everything fiendish is excluded from the geometrical construct. One example is Bonaiuto's fresco *The Apotheosis of Saint Thomas Aquinas* (1635–67) in Florence's Dominican basilica of Santa Maria Novella.[3] The cosmos pleasing to God is structured by rules and dimensions that also determine the scope of mankind's vision. Where this cosmos as perceived or controlled by man comes to an end, the world passes over into the unknown: There, man either approaches paradise or becomes lost in the undergrowth of the jungle. He leaves the sensible world behind, and strangeness and asymmetry await. The art historian Martin Warnke interpreted the impenetrable, dreamlike landscape pictures of Roelant Savery (ca. 1577–1639) as representations of such outer reaches (fig. 3).[4] Exotic pictures of rare animals and people from other continents were displayed in the galleries of the Holy Roman Emperors as reflections of the extent of their own authority, indications of his ultimate frontiers—not just idyllic scenes, but painted boundary posts as well.

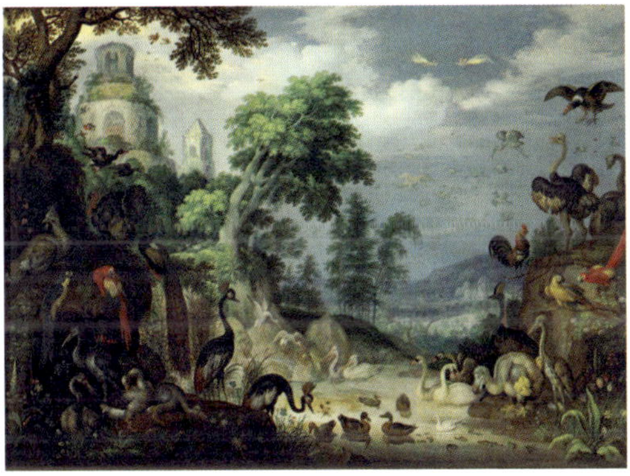

Fig. 3. Roelant Savery. *Landscape with Birds*, 1628. Copper. Kunsthistorisches Museum, Vienna. Erich Lessing / Art Resource, NY.

From the point of view of European conquerors, these paintings were invitations to extend such boundaries, to delete godless regions from the map—Emperor Charles V famously adopted the motto "*plus ultra*." This would later take the form of a battle against anarchy, a lack of leadership, and the inability to act, which are feared to this day to constitute a "power vacuum." Such emptiness cannot be because it must not be: It would be unnatural, a nothingness in the midst of the here and now.[5]

When, in the mid-seventeenth century, physicists were able to show that there is no such thing as a *horror vacui*, that vacuums are possible on earth, rulers in all countries promptly showed an interest in repeating the relevant experiments at their courts, as they wanted to regain a hold on order. They knew that it was always the fate of mankind, never wholly capable of nullifying the forces of nature, to deal with the dynamism of the unavoidable—the dynamics of nature, Fortuna, and mankind itself.

To discover the laws behind seeming lawlessness, a bigger picture is required. In early modern history, this insight gave rise to the insurance business, probability theory, cameralistics, and organization theory, as well as to experimentalism in science and art and to freedom of trade and the press.[6] All of these are expressions of an effort to maintain control by making room again for chance in the midst of order.

Productive Forces: From Nature's Creativity to the Assembly Line

Nature continues to be creative and full of surprises. The oceans and remote continents are constantly yielding up new, miraculous forms and variants that do not appear in any of our taxonomies, and strange apparitions present themselves in telescopes and microscopes.[7] How is it that such creatures can all bear a resemblance to one another and yet be distinguishable? Does nature consist of types and their variations, or is each of its creatures unique? Is nature a random generator, or does it work like a factory?

"The exception proves the rule." This conviction led to the formation of the art and curio cabinets of the seventeenth and eighteenth centuries. In them, it seemed possible to study nature as if in a laboratory.[8] They offered glimpses into the world's most secret and whimsical realms. Their collections were not merely rarities that induced amazement, they were also expressions of a scholarly curiosity that hoped to find the source code for existence in the book of nature. The search for natural laws was also based on the idea that there is a coherence behind the wealth of appearances that must be understood if it is to be tapped into. In an era of increasing industrialization, these laws constituted a step toward an inventory of nature's creations, one taken with the idea of making them productive.

For a long time, even though many discoveries radically challenged familiar rules and many boundaries gradually shifted—resulting in ever newer forms of the unknown being discovered (the chimeras and monsters of Romanticism, for example, or the world of the psychic drives)—the classification of the *systema naturae et artis* proved to be stable. It was promoted by advancements in graphic reproduction, which allowed existing knowledge to be arranged and disseminated in the form of herbals, collection catalogues, emblem books, and anatomical atlases.[9] Illustrated books adopted the basic principle of comparative, encyclopedic study from curio cabinets and art collections, offering readers numerous overview plates and picture sequences.

With these illustrations, a temporal dimension gradually came into play as well. An innovative example of the scientific application of a picture sequence was the embryological series produced by the physiologist Samuel Thomas Soemmering (1755–1830). It appears to depict the development of the human ovum into a fully formed embryo (fig. 4), yet in Soemmering's day, the human ovum had not even been discovered. His illustration was thus merely speculative: Working backward, he pictured how the embryo had to be imagined in its earliest stages. Step by step, it encroached onto divine but forbidden territory.

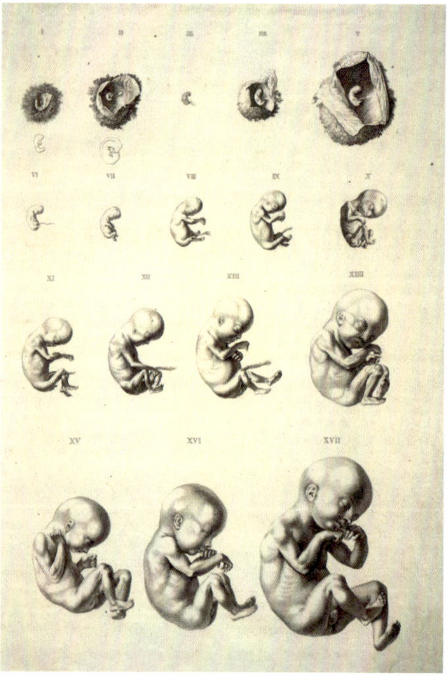

Fig. 4. Samuel Thomas Soemmering. *Icones embryonum humanorum*. Frankfurt am Main: Varrentrapp & Wenner, 1799, plate 1. Courtesy of the Huntington Library, San Marino, California.

Moreover, Soemmering's series illustrated the morphological question of how a form can dynamically transform itself, which is to say, how it can change yet remain true to itself. The individual pictures of the embryo are similar to each other, but they're not identical. They suggest a process of growth that the modern viewer registers by reading from left to right (like a sequence of film stills). In return, it is easy to forget that the plate pictured a purely hypothetical development, that it traced a rhythm of growth, at a time when mechanisms such as cell division were still unknown.[10]

Picture sequences as such were by no means novelties around 1800—narrative pictorial ensembles are found on triumphal columns, in polyptychs, in illuminations, and in woodcuts. After 1800, they were developed still further (in graphic novels, pictorial reporting, and other genres).[11] But in Soemmering's plate, the combination of natural observation and comparative pictures communicated a new sense of life, time, and growth.[12] In a sense, space and time became scientific subjects only because they could be appraised in pictures.

Pictorial representations linked objects separate in space and time in chronological or causal chains.[13] They constructed biological affinities and evolving artistic styles. Such "evolutionary" schemata proved to be as attractive to the student of natural history as to the art historian. A famous example was provided by John Ruskin (1819–1900) with his compilation and illustration of architectural forms in *The Stones of Venice*. His comparison of forms suggested the activity of higher laws and exigencies behind individual works while also indicating that there might be structural similarities between artifacts and natural phenomena. Thus the anthropologist Augustus Pitt Rivers (1827–1900), Ruskin's fellow countryman and contemporary, hypothesized in 1874,

> Human ideas, as represented by the various products of human industry, are capable of classification into genera, species, and varieties, in the same manner as the products of the vegetable and animal kingdoms, and in their development from the homogeneous to the heterogeneous they obey the same laws. If, therefore, we can obtain a sufficient number of objects to represent the suc-

cession of ideas, it will be found that they are capable of being arranged in museums upon a similar plan.[14]

Could such developmental sequences one day allow us to chart missing links in a historical development and predict the existence of finds that have not yet come to light? Here the collector remained skeptical: "Progress is like a game of dominoes—like fits on to like. In neither case can we tell beforehand what will be the ultimate figure produced by the adhesions; all we know is that the fundamental rule of the game is sequence."[15]

Nevertheless, there was still a strong hope that it might be possible to make predictions by employing comparative means (using pictures in particular) and not only with regard to the evolution of style. Humans were being turned into objects as well. Resuming the physiognomical study around 1800, and in the wake of social upheaval, urbanization, and anonymization, anthropological and criminological records began to detect specific types and situations against which society allegedly needed to be protected.[16] People's faces were made comparable by the use of standardized photographic portraits not only for the purposes of border control and law enforcement but also in order to search for indications of deviance that could affect the health of society.

The resulting files are industrialized, streamlined assembly-line products, but they are also symbolic of a craze for order to which all modern societies and political systems will be susceptible so long as they continue to equate the notions of "the state" and "order" with the preservation of the status quo.

Envisioning the Invisible

Scientific picture collections, wall charts and visual aids, and book illustrations are examples of how old schemata and representational traditions have been transformed over the centuries. Even early medieval cosmological diagrams have been revived—for example in the first *organigram* of the Scottish-American railway pioneer Daniel C. McCallum (1815–78), who used it to picture the structure of his organization in 1855.[17] Or by Walter Gropius (1883–1969), who in order to illustrate the Bauhaus curriculum resorted to motifs used in the early seventh century by Isidore of Seville (ca. 560–636).

The wealth of such forms in the world of today's visual communications is immense. They serve as intellectual tools for illustrating quantitative and power relationships, translating processes into closed loops, and suggesting contrasts and trends (before/after, worse/better).[18] In all these variants, actual lines, legends, and configurations stand for ideas and abstract situations. Conversely, they produce a reality of their own, one that reflects historical styles.[19] There are no objects, there is no nature (and also no chaos) lurking behind them; such concepts are only evoked through the use of visual media, just as through words and numbers.

In a permanent reciprocal process, these media formulate problems, and in so doing they generate them. The meaning of a diagram or a wall chart depends on its design elements, on how well or poorly it was produced. It is subject to rules of taste as well as communication. Many such graphics betray a desire to make tangible the hidden workings of the world while also demonstrating the difficulty of giving form to something that has none—including the diagram shown in figure 5, which is titled "Order and Chaos" and may serve here as one out of millions of examples to show that decisive connections are established only in the mind of the viewer who has visually read the graphic structure (and is usually lost without legends and additional explanations).

Such mental collaboration was noted by the Austrian biologist Paul Kammerer (1880–1926). He observed that there are everyday phenomena that appear to relate to each other even if causal connections can be excluded (an effect that also fascinated the Surrealists and that seems related to our sense of there being such a thing as a Zeitgeist or Murphy's Law). In his book

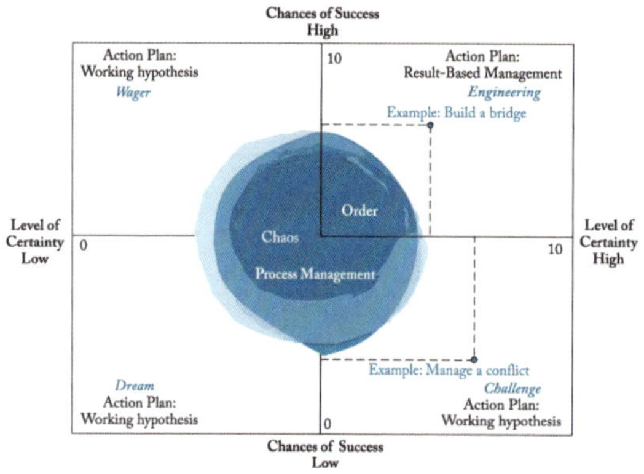

Fig. 5. "Order and Chaos." From Jacques M. Chevalier and Daniel J. Buckles, *SAS² – A Guide to Collaborative Inquiry and Social Engagement* (Los Angeles: Sage, 2009).

Das Gesetz der Serie (The law of seriality), he postulated that it is possible for one thing to necessarily follow another without there being any discernible reason for it.[20] After Kammerer came to be suspected of falsification in his experiments on the hereditary transmission of acquired traits, he took his own life.[21] His name was long in disrepute, but his book title became proverbial.

As long as science is based on observation and comparison, associative mechanisms in the beholder's mind will play a role. Repeating "patterns," average "values," and "significant" deviations are identified by employing large numbers of pictures and automated technologies, thus combining intellectual and statistical methods. Sociological studies, astronomical charts, and sequential radiological images all involve the comparison of data that can often be studied only in pictorial form. Endless sequences of pictures in electron microscopy and medical imaging are the result of a fumbling, experimental search that is based on measurements and human attention to the results of those measurements. In return, reliance on parameters and controllers becomes a component of scientific thought.

Pictures of "chaos theory" are also the synthetic product of such experimentation with variables, pencils, and plotting devices. Inasmuch as these pictures produced under human control are intended to give shape to something invisible or incomprehensible, they are once again imposing aesthetic form on raging patterns of current or the noise of electronic data. Accordingly, these shapes fail to illustrate the "chaos" flickering behind the image; they rather reveal how people think about it. Above all, images of chaos theory illustrate how the history of the visual arts has inscribed itself into such thinking.

1 For additional examples, see Petra Roettig, "Kehraus. Der Besen als politisches Bildmotiv," in *Politische Kunst*, ed. Martin Warnke (Berlin: Akademie, 2004), 33–52.
2 Klaus Wasmund, *Politische Plakate aus dem Nachkriegsdeutschland. Zwischen Kapitulation und Staatsgründung 1945–1949* (Frankfurt am Main: Fischer, 1986), 146.
3 Alexander Perrig, "Masaccio's 'Trinità' und der Sinn Der Zentralperspective," *Marburger Jahrbuch für Kunstwissenschaft* 21 (1986): 15–18. See also Erwin Panofsky, *Perspective as Symbolic Form* (1966; New York: Zone Books, 2002).
4 Martin Warnke, *Political Landscape: The Art History of Nature* (London: Reaktion Books, 1994). These motifs were further exploited in the Netherlandish forest still life (Italian: *sottobosco*) of the seventeenth century.
5 See also Edgar Wind, *Art and Anarchy* (London: Faber and Faber, 1963), who followed Aby Warburg in assuming that the creative furor in art has been progressively tamed since the age of the Renaissance.
6 For the concept of the organization, see Tobias Cheung, "From the Organism of a Body to the Body of an Organism: Occurrence and Meaning of the Word 'Organism' from the Seventeenth to the Nineteenth Centuries," *The British Journal for the History of Science* 39 (2006): 319–39; Cheung, ed., *Transitions and Borders Between Animals, Humans, and Machines, 1600–1800* (Leiden: Brill, 2008); see also Richard C. Bowler, "Mediating Creative Nature and Human Needs in Early German Political Economy," *History of Political Economy* 40 (2008).
7 Lorraine Daston and Katherine Park, *Wonders and the Order of Nature, 1150–1750* (New York: Zone Books, 1998).
8 Horst Bredekamp, *The Lure of Antiquity and the Cult of the Machine: The Kunstkammer and the Evolution of Nature, Art, and Technology* (Princeton: Markus Wiener, 1995).
9 For the scientific print in the early period, see the recent exhibition catalogue *Prints and the Pursuit of Knowledge in Early Modern Europe*, ed. Susan Dackerman (Cambridge, MA: Harvard Art Museums, 2011); for the beginnings of antiquarianism, see Francis Haskell, *History and Its Images: Art and the Interpretation of the Past* (New Haven: Yale University Press, 1993).
10 Janina Wellmann, *Die Form des Werdens: Eine Kulturgeschichte der Embryologie, 1760–1830* (Göttingen: Wallstein, 2010), 283–91.
11 For example in Rodolphe Toepffer or Hokusai, see David Kunzle, *The History of the Comic Strip: The Nineteenth Century* (Berkeley: University of California Press, 1990); also, Alan Gowans, *The Unchanging Arts: New Forms for the Traditional Functions of Art in Society* (Philadelphia: Lippincott, 1971).
12 For the importance of visual media for the evolution of geology and its conception of time, see Martin J. S. Rudwick, *Scenes from Deep Time: Early Pictorial Representations of the Prehistoric World* (Chicago: University of Chicago Press, 1992).
13 See Nick Hopwood, Simon Schaffer, and Jim Secord, ed. *Seriality and Scientific Objects in the Nineteenth Century* (Cambridge: Science History Publishing, 2010).
14 Augustus H. Lane-Fox Pitt Rivers, "On the Principles of Classification Adopted in the Arrangement of His Anthropological Collection, Now Exhibited in the Bethnal Green Museum," *Journal of the Anthropological Institute of Great Britain and Ireland* 4 (1875): 294. (I thank Stefanie Klamm, Berlin, for the reference to Pitt Rivers.)
15 Ibid., 308.
16 As described by Allan Sekula, "The Body and the Archive," in *The Body: A Reader*, ed. Mariam Fraser (London: Routledge, 2005), 163–66.
17 Alfred D. Chandler Jr., *Strategy and Structure: Chapters in the History of the American Industrial Enterprise* (Cambridge, MA: MIT Press, 1962).
18 Examples of modern design are found in Edward R. Tufte, *Beautiful Evidence* (Cheshire, CT: Graphics Press, 2007); and in the collection of "network maps" in Mark Ovenden, *Transit Maps of the World: The World's First Collection of Every Urban Train Map on Earth* (London: Penguin, 2007). See also the impressive example by Heinrich von Geymüller, analyzed in Karsten Heck, "Der Schreibtisch als Denkraum. Heinrich von Geymüllers 'Tableau graphique,'" in *Heinrich von Geymüller (1839–1909)*, ed. Josef Ploder and Georg Germann (Basel: Reinhardt, 2009); and Alfred H. Barr's "Genealogy of Art" in Astrit Schmidt-Burckhardt, "Shaping Modernism: Alfred Barr's Genealogy of Art," *Word & Image* 16 (2000): 387–400.
19 See especially Caroline A. Jones and Peter Galison, eds., *Picturing Science, Producing Art* (New York: Routledge, 1998); also Horst Bredekamp, Birgit Schneider, and Vera Dünkel, eds., *Das Technische Bild. Kompendium zu einer Stilgeschichte wissenschaftlicher Bilder* (Berlin: Akademie, 2008).
20 Paul Kammerer, *Das Gesetz der Serie. Eine Lehre von den Wiederholungen im Lebens- und Weltgeschehen* (Stuttgart: Deutsche Verlags-Anstalt, 1919).
21 Kammerer's fall has been recently discussed in Pennisi, "The Case of the Midwife Toad: Fraud or Epigenetics?," *Science* 325 (2009), and Alexander O. Vargas, "Did Paul Kammerer Discover Epigenetic Inheritance? A Modern Look at the Controversial Midwife Toad Experiments," *Journal of Experimental Zoology* 312 B (2009).

Found Images II

Order, Disorder, Beauty

Alan Norton
with
Nina Samuel

I wanted it to have it look like a majestic old-fashioned dress. This picture was made to show off. It was for a computer journal. I mean it's like a fashion show. If you want to make a point by pictures you use every trick in the game. And the game is the same whether it's a fashion show for dresses or whether it's a fashion show for such shapes. It was an illustration which has not led to any mathematical discovery. It was part of this feeling: The tools exist; it can be done; let's do it![1] (cat.1) —Benoît Mandelbrot

The images in this section were programmed by Alan Norton. He recalls his time with Mandelbrot and explains how and why these images were made:

> I worked with Mandelbrot as a postdoc for two years starting in July 1980. My job was mostly to provide computer illustrations for *The Fractal Geometry of Nature*. I made a variety of images based on Benoît's interests. He showed me how one could use the Mandelbrot set to find Julia sets (dragon-like images identified by the early-twentieth-century mathematician Gaston Julia without the benefit of computer graphics) that have various properties, such as twisting spirals or separated components.
>
> As a mathematician, I was struck by the idea that every formula has a corresponding picture and that these pictures are usually fractals. My background in 3D computer graphics led me to wonder if these stunning shapes could be viewed in three or more dimensions. I knew that any of the formulas defining Julia sets might also be used to define higher-dimensional shapes. For example, the nineteenth-century physicist William R. Hamilton had discovered a four-dimensional space, the "quaternions," to which these dragon shapes naturally extended. My first effort was to imagine what these extensions might look like by calculating quaternion Julia sets on two-dimensional planes that extended into the third dimension. This resulted in odd images (cats. 2–5). Benoît and I discussed these odd planar shapes that I was generating, and we agreed that this was an interesting research direction. He supported my proposed effort to develop the software to make pictures of Julia sets in three dimensions.
>
> When I started making these 3D images, I was very excited to find a multitude of different shapes that I never

could have imagined. They were wholly unlike the originally planar Julia sets. Benoît too was intrigued by the quaternion Julia sets.

Neither of us could understand the complex looping patterns, which appeared to be almost accidental. I puzzled over that problem for a year or more, making hundreds of 3D images in the process. Eventually, I found a simple rule that could be used to determine how the loops in the quaternion Julia sets were connected in the third dimension. This enabled me to deliberately design different connection patterns, such as the ones found in the images reproduced here (cats. 6–9).[2]

In this field of study scientists and programmers tried to make two-dimensional images look like "real things": shaded, tangible, and with a distinct texture, ultimately suggesting they could be manipulated and touched like objects in the real world. This impulse was misleading since, strictly speaking, these images represented nothing else than an array of numbers in a plane. The capability of these computer-generated images was enthusiastically exploited for images that were produced for popular publications. With the help of Alan Norton, Mandelbrot took decisive steps towards decorative designs that would capture the imagination of audiences for decades to follow and were largely responsible for the popular idea "how fractals look." However, as emphasized in the epigraph above, the gap between these popular "illustrations" and a mathematical discovery through images, is immense.

1 Benoît Mandelbrot, interview by Nina Samuel, Cambridge, MA, April 2008.
2 Alan Norton, e-mail to Nina Samuel, March 13, 2012.

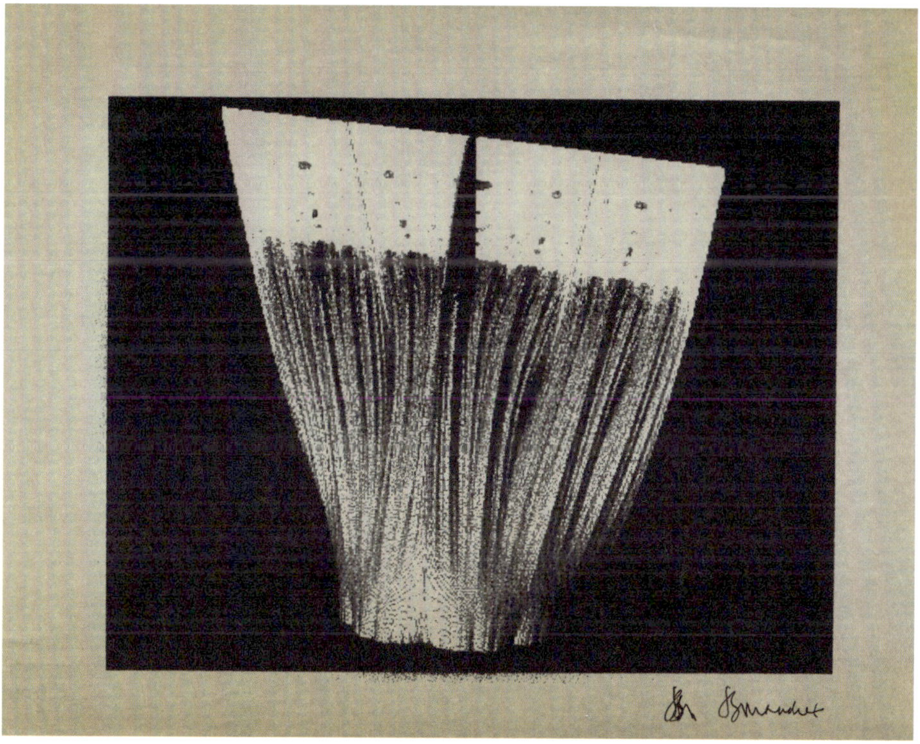

Cat. 1. Benoît Mandelbrot and Alan Norton, programmer. *Quaternion*, ca. 1982. Computer-generated print. Collection Aliette Mandelbrot.

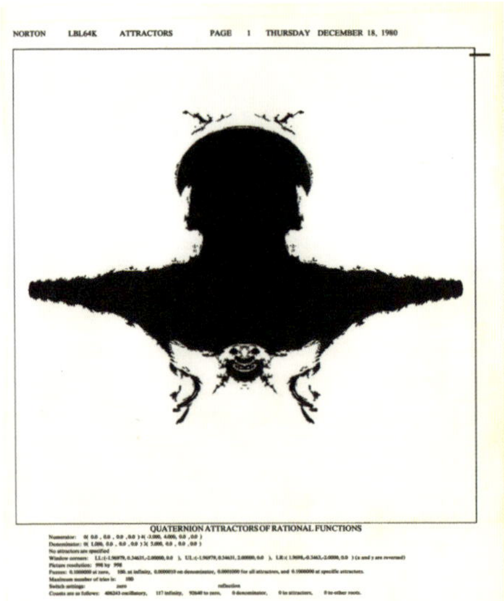

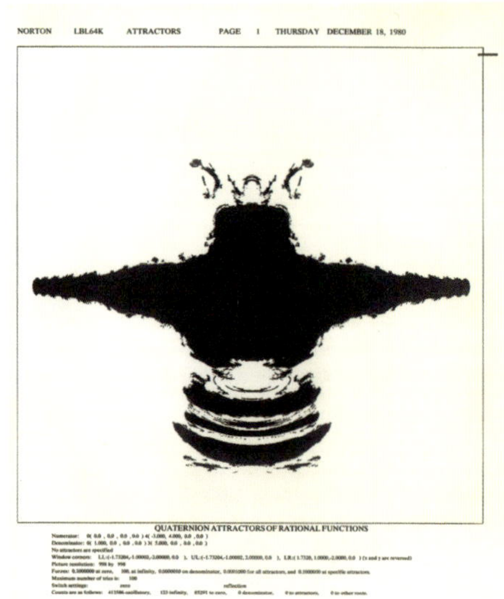

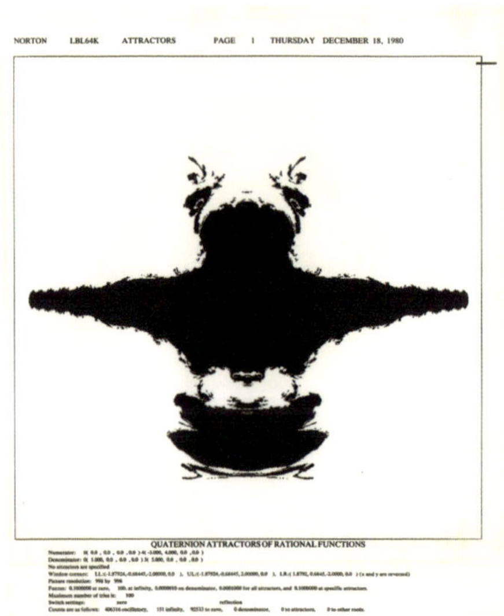

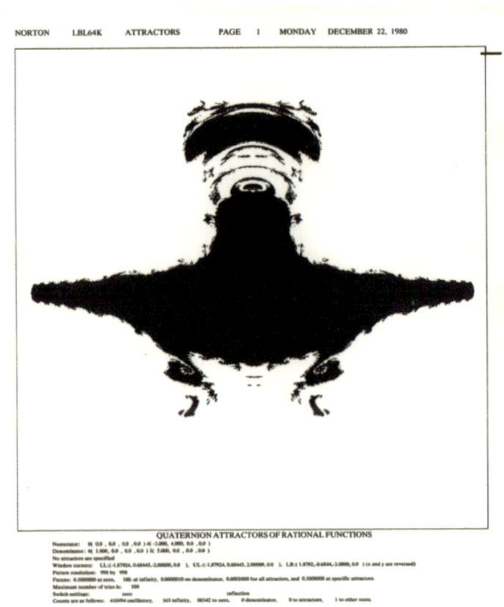

Cats. 2–5. Benoît Mandelbrot and Alan Norton, programmer. *Quaternion*, 1980. Computer-generated prints on photographic paper. Collection Aliette Mandelbrot.

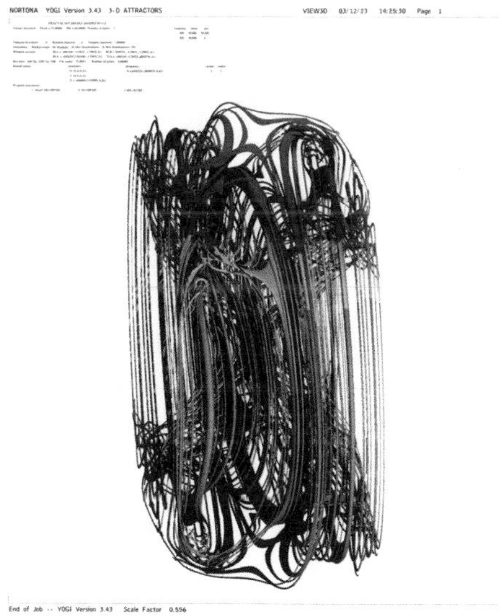

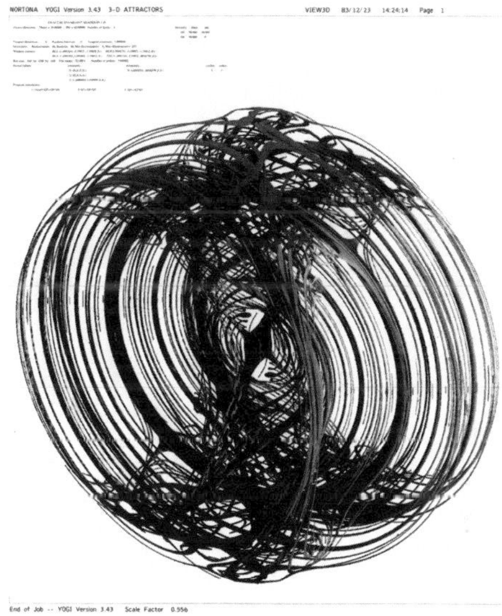

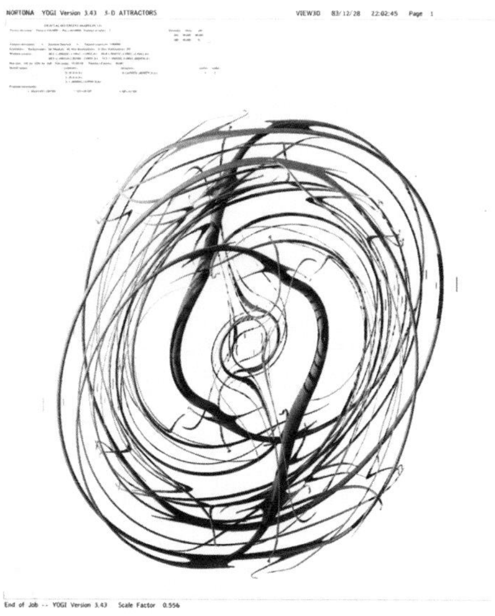

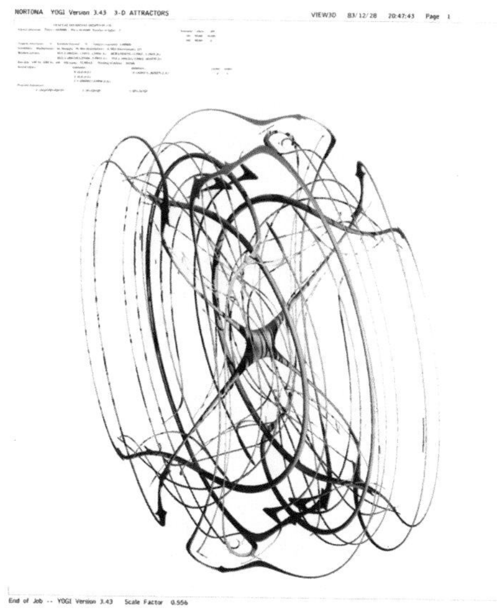

Cats. 6–9. Benoît Mandelbrot and Alan Norton, programmer. *Quaternion*, 1983. Computer-generated prints on photographic paper. Collection Aliette Mandelbrot.

Seeing Order in Disorder

Margarete Pratschke

"I am afraid my answer will hardly satisfy you":[1] With these words, the art historian Ernst Gombrich (1909–2001) attempted to fend off a question from the art psychologist Rudolf Arnheim (1904–2007) about the relationship between order and disorder. With the courtesy appropriate to a conversation between colleagues, Gombrich made it clear that despite their fundamental agreement on the relationship between visual perception and knowledge, he could not go along with Arnheim on at least one point: When it came to explaining how their basic agreement related to visual order and disorder, Gombrich placed far greater importance on disorder than Arnheim. But what might have seemed dismissible as a minor disagreement was in fact the result of their diametrically opposing points of view regarding the way visual perception and knowledge are related, based as they were on altogether different theoretical premises.

The occasion for Arnheim and Gombrich's discussion of visual order and disorder was a 1986 symposium and exhibition on symmetry at which the recently emerged fascination with fractals was a topic under review. Fractals were being celebrated for the way they visualize mathematical "chaos" as formal order and thus seem to bring the dichotomy between order and disorder into harmony.[2] A mathematician from the field of chaos studies, Heinz-Otto Peitgen, took part in the discussion as well. A vocal exponent of the aesthetic quality of fractals, he considered them worthy subjects for interdisciplinary discourse—a useful vehicle for making the kind of formal comparisons between science and art that could be found in the exhibition (fig. 1).[3]

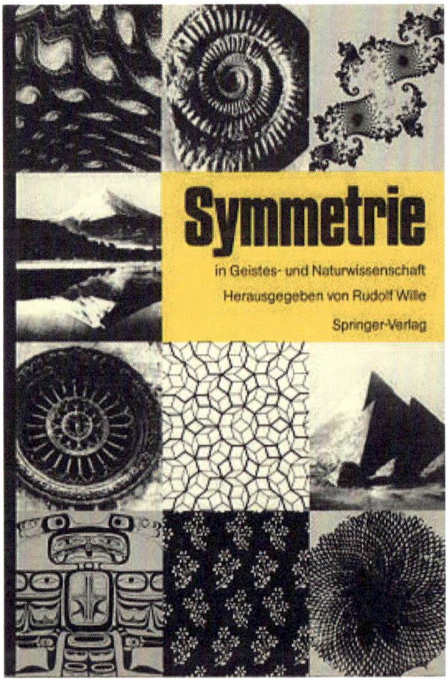

Fig. 1. Rudolf Wille. ***Symmetrie in Geistes- und Naturwissenschaft: Hauptvorträge und Diskussionen des Symmetrie Symposions an der Technischen Hochschule Darmstadt vom 13. bis 17. Juni 1986 im Rahmen des Symmetrieprojektes der Stadt Darmstadt*** **(Berlin: Springer-Verlag, 1988), cover. With kind permission of Springer Sciences and Business Media.**

How little time it took for fractals to become aestheticized can be seen from a glance at the publications of Benoît Mandelbrot. In the introduction to his 1982 book *The Fractal Geometry of Nature*, Mandelbrot states, as a kind of motto: "Fractal geometry reveals that some of the most austerely formal chapters of mathematics had a hidden face: a world of pure plastic beauty unsuspected until now."[4] Here, at the beginning of his book, he self-confidently celebrates the beauty of the picture worlds he has produced, yet only a few years before, in his preceding text from 1977, for example, there were only isolated references to the aesthetic qualities of his visualizations. In 1982, however, Mandelbrot, surely a master of self-promotion, not only speaks of beauty in connection with his computer-generated "patterns," but claims quite explicitly: "I hear and read of . . . [my] illustrations being described as a 'New Form of Art.'"[5]

To be sure, Mandelbrot's scientific pictures were indeed "new;" however, scientists' attempts to elevate their pictures to the status of "art" and to bridge the divide between art and science were not new at all. Since the 1950s, specific picture types such as patterns, grids, and abstract geometrical visualizations from the sciences, most of them mechanically produced, had evolved to suggest the close proximity between science and art and to underscore the beauty of their form. To take one example, in the foreword to an anthology produced for the exhibition "Growth and Form" at London's Institute of Contemporary Arts in 1951, Herbert Read wrote: "The increasing significance given to form or pattern in various branches of science has suggested the possibility of a certain parallelism, if not identity, in the structures of natural phenomena and of authentic works of art."[6] But whereas Read was relatively cautious in pointing to such parallels, the volume's editor, Lancelot Law Whyte, formulated the claim to interdisciplinarity with far more confidence: "Common to the ideas of form, configuration, pattern, and structure, is the notion of an ordered complexity, a multiplicity which is governed by some unifying principle."[7]

Like Whyte, György Kepes (1906–2001), an artist and the founder of the Center for Advanced Visual Studies at MIT, was committed to the search for an underlying order, making it the focus of his ambitious didactic efforts to bring form in science and art into harmony. This is perfectly apparent in his publications *The New Landscape* (1951), *Structure in Art and in Science* (1965), and *Module, Proportion, Symmetry, Rhythm* (1966).[8] In them, Kepes illustrated the "new landscapes" emerging from scientific laboratories, all of which picture "structures," which is to say geometric-abstract forms, and all of which were created by such diverse methods as aerial photography, microphotography, X-ray photography, sound visualization, oscillography, schlieren photography, et cetera (figs. 2 and 3). By juxtaposing such scientific-technical forms with works of art, he hoped to demonstrate that the creation of form in both science and art was based on a common principle: "Image-making is basic for art. It is basic for science, too. . . . On the image-making level, the difference between pre-scientific and pre-artistic perception of order is a difference of attitude. . . . Structures can be understood . . . in a single, balanced perception of order, in an

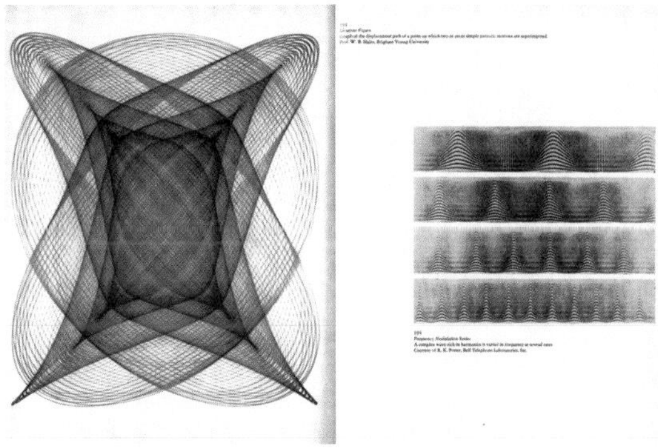

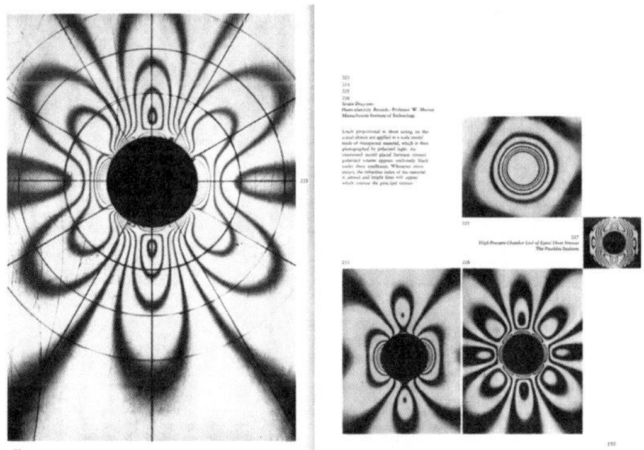

Fig. 2. György Kepes. Scientific illustrations. From *The New Landscape in Art and Science* (Chicago: P. Theobald, 1956), 176–77. Courtesy Juliet K. Stone.
Fig. 3. György Kepes. Scientific illustrations. From *The New Landscape in Art and Science* (Chicago: P. Theobald, 1956), 192–93. Courtesy Juliet K. Stone.

experience which has characteristics of scientific and artistic activity both."[9] Although Kepes's argument regarding the epistemic meaning of pictures was highly sophisticated, what was especially apparent from the illustrations in his books was his simple fascination with the aesthetics of the picture worlds then being created in the sciences in ever-increasing numbers.

Kepes's specific interest in scientific picture worlds, combined with his artistic agenda, was not without its critics. Again and again, objections were raised against both his comparing art and science and his ways of analyzing and communicating (or popularizing) scientific pictures. A reviewer of *The New Landscape*, for example, noted that Kepes had succumbed to the artist's typical fascination with analogies between the forms of art and those of nature,[10] similarities that in many cases were only coincidental.[11] The review then proceeded to voice a fundamental issue, namely that the meaning of a scientific illustration had nothing to do with its aesthetic quality but was based on wholly different criteria: "The object receives its meaning from qualities unrelated to its form. For every scientific picture presented in this book because it is accidentally beautiful there were omitted multitudes which equally accidentally were ugly or inexpressive, but whose significance is undiminished by this esthetic shortcoming." Distinctly unattractive pictures could be extremely meaningful, the reviewer suggested, in their experimental, scientific, functional context, a context that Kepes had ignored. The review concluded with the point that a layman presented with a scientific picture by itself could by no means guess its experimental context and thus its meaning: "The total ignorance of the layman concerning the forces that produced this landscape separates it once and for all from art."[12]

What put critics off was not so much the question of whether these pictures were art or not, but how one might properly analyze them outside of the context in which they were created. This involved the fundamental question of what function one might accord to pictures in science in general. Many writers saw in them only "visual by-products of scientific thoughts";[13] for others, their visibility was precisely the key to understanding them. The premise upon which Kepes based his overall claim of a connection between art and science was his conviction that seeing brought with it knowledge: "Vision is by no means an automatic function of our physiological apparatus. There is much evidence that vision is itself a mode of thinking."[14] He felt that by explaining how seeing was to be conceived as thinking, and how visual perception could be accorded an epistemic function, he was contributing to the understanding of scientific picture structures and their analysis.

But here was precisely where the thinking of the art historians Gombrich and Arnheim diverged. From their knowledge of the psychology of perception, art history, and scientific theory, they had come to different conclusions about how visual thinking was to be explained with respect to abstract form or pictorial structures, and for both of them the relationship between visual order and disorder, the point on which they disagreed, played a central role.

Perhaps more insistently than anyone else in the twentieth century, Rudolf Arnheim argued that the act of seeing involves a cognitive process. His *Visual Thinking* (1969) was a philippic against epistemologies based on language; these first arguments had already appeared in 1954's *Art and Visual Perception: A Psychology of the Creative Eye*.[15] There he stressed the fact that "far from being a mechanical recording of sensory elements, vision turned out to be a truly creative grasp of reality—imaginative, inventive, shrewd, and beautiful. . . . All perceiving is also thinking, all reasoning is also intuition, all observation is also invention."[16] Arnheim was referring explicitly to Gestalt theory, which had demonstrated in countless experiments that "the appearance of any element depends on its place and function in the pattern as a whole."[17] Basic to his theory was the idea that a person's ability to infer parts from a whole pointed to the intelligence of perception (again in accordance with Gestalt theory)—to its "active striving for unity and order," which was "manifest in

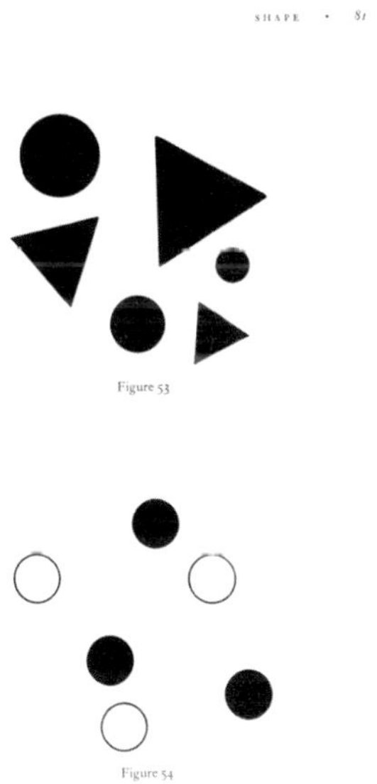

Fig. 4. Rudolf Arnheim. Illustration to explain the visual perception of similarities and differences using the example of group building and subdivision on the basis of specific features. From *Art and Visual Perception: A Psychology of the Creative Eye* (Berkeley: University of California Press, 1954), 81.

the simple act of looking at a simple pattern of lines" (fig. 4).[18] For Arnheim, seeing searches for simplicity and order; it resolves tension within the visual field and changes disorder into a fundamental order of perceptual impressions. He thought of this process not only as a kind of direct visual problem solving but also as an epistemic procedure in which "thought operations of discovery and invention"[19] were made manifest. Arnheim saw the striving for order, which he always preferred over disorder, as wholly confirmed by chaos theory and the visualizations of fractals:

> In many areas of science and art, basic structural constituents can be discerned even in complex manifestations. Recent explorations of "chaos" have shown that . . . natural phenomena considered totally irrational can be reduced to simple mechanisms. This is true for the patterns combining repetition, growth and randomness in the theory of natural fractals.[20]

Gombrich also thought of patterns as being essential to a visual theory of knowledge. With his 1979 book *The Sense of Order: A Study in the Psychology of Decorative Art*, the notoriously harsh critic of modern art developed nothing less than an epistemology of abstract form.[21] Frequently dismissed as a book simply about ornament, one overshadowed by his main work, *Art and Illusion*,[22] *The Sense of Order* contains a comprehensive theory of repeating patterns, which Gombrich argues occupy a central function in perception. He was less interested in decorative patterns as such than in a universal visual principle that he characterized as "the arrangement of elements according to similarity and difference . . . repetition and symmetry."[23] Moreover, he assumed that the use of simple geometrical forms in the world created by mankind was an expression of a "sense of order."[24] Wholly in line with the notions of Gestalt theory, which were shared by Arnheim and Kepes, Gombrich emphasized the constructive, active aspect of perception. He viewed the "organism as an active agent reaching out towards the environment, not blindly and at random, but guided by its inbuilt sense of order."[25] Initially borrowing from Gestalt theory as well, Gombrich proposed that "there is an observable bias in our perception for simple configurations, straight lines, circles and other simple orders and we will tend to see such regularities rather than random shapes in our encounter with the chaotic world outside."[26]

However, at the same time, Gombrich did indeed limit the predominance of this fundamental order, shifting his focus to disorder as a way of explaining the emergence of novelty, or knowledge.[27] To him, it was deviation that offered the potential for discoveries.[28] With this shift in focus, Gombrich quite openly turned away from Gestalt theory to information theory as a referential model—that is to say, to a cybernetic model that accorded an important role to "disorder." The advantage of the information-theory model, according to Gombrich, lay in the way it measured information by "its degree of unexpectedness, while the expected becomes... the 'redundant.'"[29] It was in this way that Gombrich judged the inherent cognitive potential of ornamental forms or patterns: "the regular row of dots, once defined and matched, merely repeats and brings nothing new."[30] Only breaks or interruptions—in other words, *disorder*—might provide something new, which is to say, specific information.

With its reference to cybernetic models, Gombrich's explanation of the perception of abstract form diverged greatly from those of Gestalt theory and Arnheim. Whereas in Gestalt theory, reasoning proceeds from the whole to the parts, or from the top down, in cybernetics, which is oriented toward computer operations, it proceeds in linear fashion, additively, piece by piece, from the parts to the whole. Insofar as one placed greater importance on order and the other on disorder, these two men found themselves in different ideological camps that could not have been more opposed in their positions vis-à-vis the computer.[31] Whereas Arnheim rejected the computer and explanations of visual-cognition processes derived from information theory, Gombrich not only used this theory, but quite practically welcomed "an exactly calculated dose of randomness" from the computer for the production of patterns.[32] With reference to an example from the artist Georg Nees (fig. 5), Gombrich expressed his hope for computer graphics: "The configurations resulting from this technique of giving latitude to chance within a fixed framework of predetermined moves not only exhibit features of aesthetic interest, they also offer unrivalled insights into the operations of our sense of order in the perception of complex patterns."[33] This was not about aesthetics, but about visual thinking, the sense of order one experiences when faced with complex patterns, which mathematics was also producing with the help of the computer.

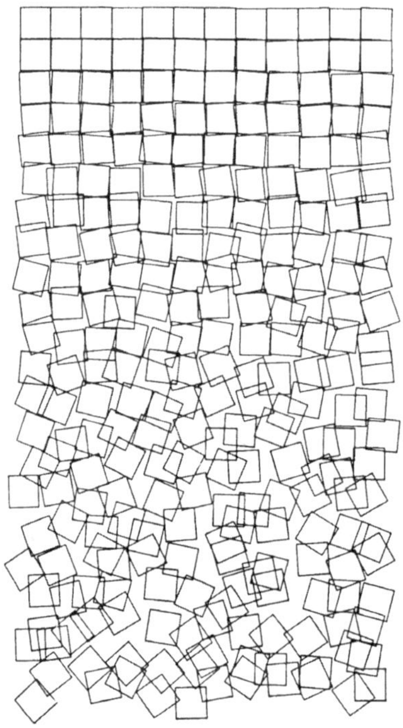

Fig. 5. Georg Nees. *Gravel Stones*, 1968–70. Lithograph on paper. © Victoria and Albert Museum, London. Given by the Computer Arts Society, supported by System Simulation Ltd., London.

When Arnheim and Gombrich clashed in 1986, it appeared that a long-simmering debate had broken out once again on the topic of fractals. At the core of it was the question of how visual knowledge in the sciences might be appropriately analyzed. As irreconcilable as the two art historians' approaches were in this regard, they were wholly in accord on one point: the relationship between art and science with respect to scientific pictures, which were all too readily being declared to be art. Arnheim was totally on Gombrich's side when he assured the latter that his occupation with non-artistic images was not about showing "that the particular quality we call art can be found also in the more secular varieties of pictures but rather to point out that nothing essential distinguishes works of art from pictorial representation for other purposes."[34] What was essential was how pictures function, how they move between perception and knowledge. Arnheim even allowed himself to characterize Gombrich's engagement in this matter as an "ecumenical initiative."[35]

1 In Rudolf Wille, ed., "Diskussion zu den Vorträgen von Sir Ernst Gombrich, Frei Otto and István Hargittai in Verbindung mit dem Thema 'Ordnung und Orientierung durch Symmetrie,'" in *Symmetrie in Geistes- und Naturwissenschaft: Hauptvorträge und Diskussionen des Symmetrie-Symposiums an der Technischen Hochschule Darmstadt vom 13. bis 17. Juni 1986 im Rahmen des Symmetrieprojektes der Stadt Darmstadt* (Berlin: Springer, 1988), 162.
2 Bernd Krimmel, ed. *Symmetrie in Kunst, Natur und Wissenschaft*, exh. cat. (Darmstadt: Mathildenhöhe, 1986); see also Wille, *Symmetrie in Geistes- und Naturwissenschaft*, as well as the report in Karl Clausberg, "Feigenbaum und Mandelbrot. Neue Symmetrien zwischen Kunst und Naturwissenschaften," *Kunstforum International* 85 (1986).
3 Heinz-Otto Peitgen and Peter Richter, *The Beauty of Fractals: Images of Complex Dynamic Systems* (Berlin: Springer, 1986).
4 Benoît Mandelbrot, *The Fractal Geometry of Nature* (New York: Freeman, 1982), 4.
5 Ibid., 23. However, it seemed important to Mandelbrot to immediately add a caveat: "Clearly, competing with artists is not at all the purpose of this Essay."
6 Herbert Read, "Preface," in *Aspects of Form: A Symposium on Form in Nature and Art*, ed. Lancelot Law Whyte (London and Bradford: Lund Humphries, 1951), v.
7 Lancelot Law Whyte, "Introduction," in *Aspects of Form* (London: Lund Humphries, 1951), 2.
8 György Kepes, *The New Landscape in Art and Science* (Chicago: Paul Theobald, 1956); Kepes, ed., *Structure in Art and in Science* (New York: Braziller, 1965); and Kepes, ed., *Module, Proportion, Symmetry, Rhythm* (New York: Braziller, 1966).
9 Kepes, *The New Landscape in Art and Science*, 22.
10 "The collection conveys the wonder with which a truly artistic eye perceives and enjoys the richness of forms and patterns now open to the senses through new recording methods." Sibyl Moholy-Nagy, "Review: The New Landscape in Art and Science by György Kepes," *College Art Journal* 18 (1959): 192.
11 Ibid.
12 Ibid., 194.
13 Richard Hamilton, "Review: The Visual Arts Today, ed. György Kepes," *Design* 154 (October 1961): 91.
14 Kepes, *The New Landscape in Art and Science*, 17.
15 Rudolf Arnheim, *Art and Visual Perception: A Psychology of the Creative Eye* (Berkeley: University of California Press, 1954); Arnheim, *Visual Thinking* (Berkeley: University of California Press, 1969). For the question of visual order, see also Arnheim, *Entropy and Art: An Essay on Disorder and Order* (Berkeley: University of California Press, 1971); Arnheim, "From Chaos to Wholeness," *The Journal of Aesthetics and Art Criticism* 54 (1996).
16 Arnheim, *Art and Visual Perception*, viii.
17 Ibid., vii.
18 Ibid., viii.
19 Arnheim, *Visual Thinking*, vi. There he writes in detail: "productive thinking in philosophy or science consists in shaping of images, one may seem to cling naively to the primitive beginnings of human reasoning, when theories were derived from the sensory form of what was perceived or imagined. But although there may be a difference in principle between those early explorations of nature and the techniques of processing data in our time, this difference may not be relevant for the crucial thought operations of discovery and invention."
20 Rudolf Arnheim, "Symmetry and the Organization of Form: A Review Article," *Leonardo* 21 (1988): 275. See also a conversation with Paul Rand, in which he remarks, apropos of chaos theory: "We don't like chaos"; see Paul Rand and Rudolf Arnheim, "Dialogue," in *Rudolf Arnheim: Revealing Vision*, ed. Kent Kleinman and Leslie Van Duzer (Ann Arbor: University of Michigan Press, 1996), 76.
21 Ernst H. Gombrich, *The Sense of Order: A Study in the Psychology of Decorative Art* (Oxford: Phaidon, 1979).
22 Ernst H. Gombrich, *Art and Illusion: A Study in the Psychology of Pictorial Representation* (Princeton: Princeton University Press, 1960).
23 Gombrich, *The Sense of Order*, x.
24 "The decorative shapes and patterns . . . testify to man's pleasure in exercising the sense of order by making and contemplating simple configurations regardless of their reference to the natural world. The world which man has made for himself is, as a rule, a world of simple geometric shapes." Ibid., v.
25 Ibid., 5. Drawing on the theories of the epistemologist Karl Popper and the cognition theorist Richard Gregory, Gombrich thus differed from stimulus-response theorists.
26 Ibid., 4.
27 "Thus while the Gestalt approach fastens on our perception of order I would draw attention to the reverse, our response to disorder." Ibid., 121.
28 "It appeals to the use we can make of order for the discovery of deviations." Ibid., 4. This is wholly in conformity with Karl Popper's notion of falsifiability in science, which Gombrich had already cited in *Art and Illusion* with regard to the stylistic development of representative pictures.
29 Ibid., 9.
30 Ibid., 123.
31 For the relationship between cybernetics and Gestalt theory, especially the different explanations of perception presented and debated by the cyberneticist Warren S. McCulloch and the Gestalt theorist Wolfgang Köhler, see Steve J. Heims, *Constructing a Social Science for Postwar America: The Cybernetics Group, 1946–1953* (Cambridge: Cambridge University Press, 1993), 204, 224. See also Margarete Pratschke, "Windows als Tableau. Zur Bildgeschichte grafischer Benutzeroberflächen" (PhD diss., Humbolt-Universität, Berlin, 2010).
32 Gombrich, *The Sense of Order*, 93–94.
33 Ibid.
34 Rudolf Arnheim, "Perceiving and Portraying: E. H. Gombrich, the Image and the Eye: Further Studies in the Psychology of Pictorial representation," *Times Literary Supplement*, October 29, 1982, 1179.
35 Ibid.

Found Images III

Images of Reality

Nina Samuel

Graphics is a wonderful tool for matching models with reality.[1]
—Benoît Mandelbrot

Even though the pictures looked like old-fashioned worn photographs, they looked like mountains. Having seen them, no one could say that I was barking up the wrong tree. Their eyes convinced them.[2]
—Benoît Mandelbrot

In his first book, *Les objets fractals* (1975), Benoît Mandelbrot published two collages made from the series of Polaroids reproduced here (cats. 1–8). He called them "*quelques vues d'un continent imaginaire* [some views of an imaginary continent]." He explained that for anyone who loved geometry, the best test for determining the extent to which a mathematical formula corresponded in some way to nature was the impression that passed directly to the brain from the eye. Only this, Mandelbrot contended, would allow one to distinguish various "degrees of reality."[3] These different square "slices" of a mountainous landscape actually represent a section of a computer algorithm designed to generate random movement. Mandelbrot made mountains in countless numbers, searching for mathematical parameters that would generate what he saw as an accurate image of nature (compare also catalogue illustrations 9 to 12 for the later color versions produced in collaboration with Richard Voss). Study of the wildly diverse surface structures of his "imaginary continents"—at times sharply jagged, at other times nearly flat—makes it clear that only certain values for parameter D would make them look "natural" or "realistic." Needless to say, his picture selection was preceded by an extensive series of visual tests.

In Mandelbrot's visual tests, even the characteristics of vector graphics (the particular pictorial technique he employed) were interpreted in terms of their degree of naturalness: "As luck would have it, the graphic process spontaneously created the impression that the sea is shimmering toward the horizon" (cats. 1–8).[4] The x-axis, up to which the computer had calculated the structures, was reminiscent of a variously rising or falling sea surface on which the sunlight was reflected—an effect that was more the result of the deficient graphic resolution of the network of lines in the background plane than proof of a similarity to "nature." The picture technique could have been used just as well to produce "unnatural" pictures; this is evident in the bottom corners at the back of the rectangular solids in some images, their artificial shape, reminiscent of toy blocks, is emphasized all the more.

As in any experimental process, errors emerged (compare cats. 14 and 16). An algorithmic miscalculation or software bug revealed an essential tension between Mandelbrot's computer-generated mountains and "real" nature: Though his fractal landscapes could simulate natural forms through random generation, even their randomness had to be carefully constructed and controlled to avoid failures.

While Mandelbrot's mountains only existed as mathematical formulas, he also collected images that were drawn from measurements of actual phenomena (cat. 15). Produced by a branch of the U.S. National Oceanic and Atmospheric Administration, one such image presents infrared data collected by satellites as Tropical Storm Naomi passed over Mexico in October 1976. The wireframe "landscape" is based on the storm system's temperature: Colder points appear taller, mimicking the structure of the clouds. How the image ended up in Mandelbrot's files is uncertain, but given his long-running interest in mountains, clouds, and astronomy, it is not difficult to imagine why he kept it. The image might have appeared to him as evidence of the underlying fractal geometry of nature—or he might have used it as a model to help anchor his algorithms in the natural world.

Mandelbrot had made the Polaroids (see cats. 1–8) with the help of programmer Sigmund Handelman, who worked with him from September 1974 to February 1977. A few of them were obviously trimmed with scissors: Was Mandelbrot here trying to expose volumes? Although this is difficult to determine, two especially striking finds from his office show that Mandelbrot and Handelman were definitely experimenting at the same time with a "fractal dimension D" in three-dimensional space.

The quality of "roughness" that Mandelbrot was investigating was now supposed to be capable of being "grasped" haptically, and to that end they experimented with sculpture, producing metal, block-like objects with the help of milling machines at IBM (cats. 17 and 18). Handelman describes the notion of the translation into sculpture in a pragmatic way: "At the same time, we figured out how we could carry it around. Before we went into color and the splashy effects, it was very nice to say you were working on fractals and have the mountain standing on your desk."[5] The aluminum block with an irregular surface structure stands at the beginning of a long history of decorative objects that fascination with fractals has produced. Such objects also document the delight scientists take in play, which has to be considered a driving force in research. Inasmuch as their 3D models and films were also used for didactic presentations in lectures, they can be linked to the nineteenth-century tradition of using mathematical models as teaching aids in university instruction. It was believed that being able to feel curves, planes, and functions gave students a better understanding of them.[6]

The idea that looking at fractal nature leads to a "loss of innocence" is a classic topos in the history of its reception:

> Fractal geometry will make you see things differently. There is danger in reading further. You risk the loss of your childhood vision of clouds, forests, galaxies, leaves, feathers, flowers, rocks, mountains, torrents of water, carpets, bricks, and much else besides. Never again will your interpretations of these things be quite the same.[7]

How fractal geometry could have been so seductive that writers of mathematical handbooks felt it necessary to provide "warnings" is closely related to the image of "nature" in fractal pictures.

I would like to thank Colin Fanning for his research on cats. 14–16 and his contributions to this text. I am also grateful to Nynne Christoffersen for her help with the Polaroids.

1 Benoît Mandelbrot, *Fractals: Form, Chance, and Dimension* (San Francisco: Freeman, 1977), 24.
2 Benoît Mandelbrot, quoted from Monte Davis, "Profile of Benoît B. Mandelbrot," *Omni Magazine* (February 5, 1984). Accessible under http://www.math.yale.edu/mandelbrot/interviews.html (accessed April 15, 2011).
3 Benoît Mandelbrot, *Les objets fractals: forme, hasard et dimension* (Paris: Flammarion, 1975), 112.
4 Mandelbrot, *Fractals*, 221.
5 Sigmund Handelman, telephone interview by author, January 21, 2009.
6 See Gerd Fischer, ed., *Mathematische Modelle. Aus den Sammlungen von Universitäten und Museen*, vol. 1, *Bildband*, vol. 2, *Kommentarband* (Berlin: Akademie-Verlag, 1986). See also Herbert Mehrtens, "Mathematical Models," in *Models: The Third Dimension of Science*, ed. Soraya de Chadarevian and Nick Hopwood (Palo Alto: Stanford University Press, 2004).
7 Michael F. Barnsley, *Fractals Everywhere* (Boston: Academic Press, 1988), 1.

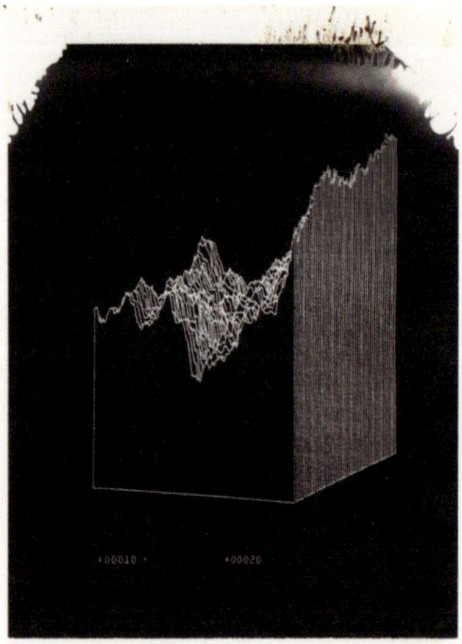

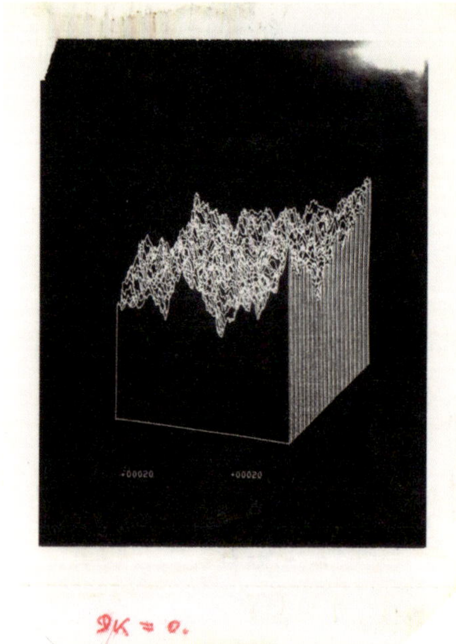

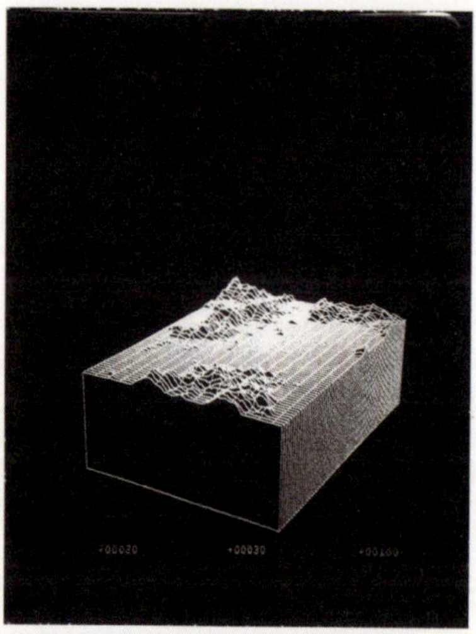

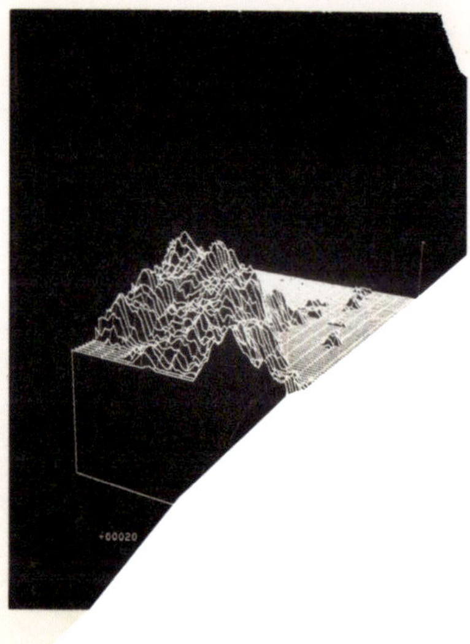

Cats. 1–8. Benoît Mandelbrot and Sigmund Handelman, programmer. *Landscapes and Galaxies*, series of forty-five, 1974–77. Polaroids. Collection Aliette Mandelbrot.

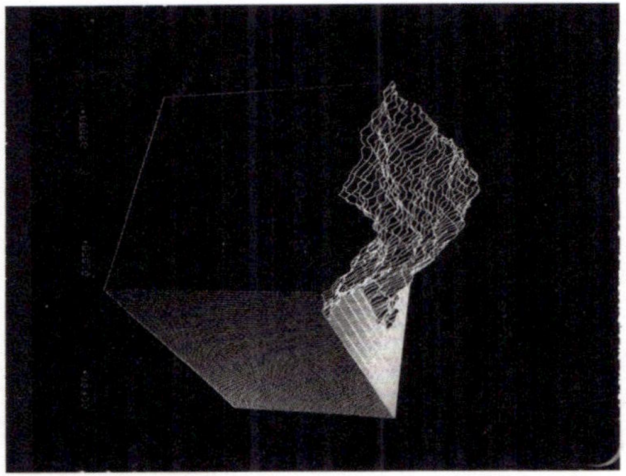

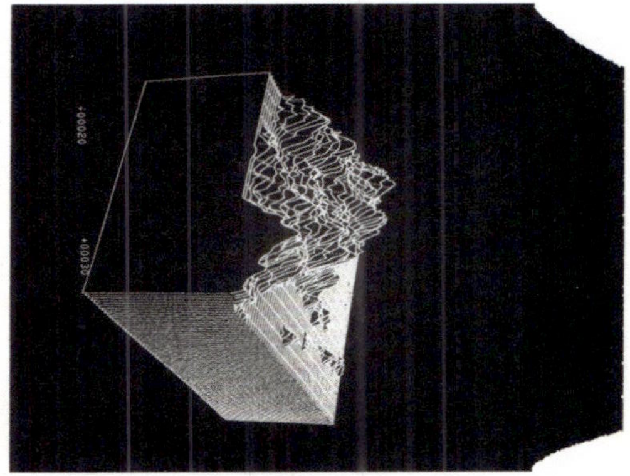
+00020
+00030

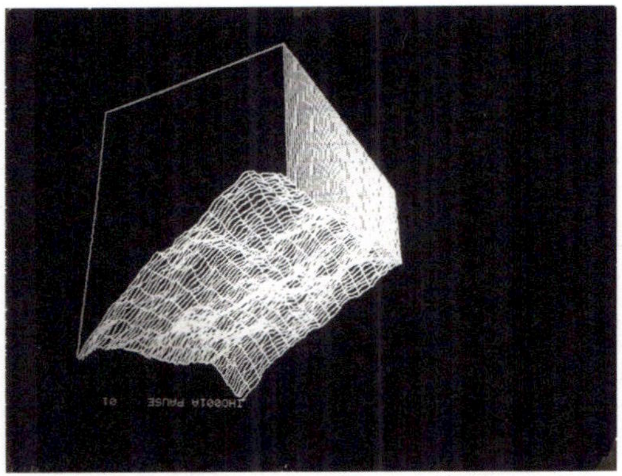
IHD001A PAUSE 01

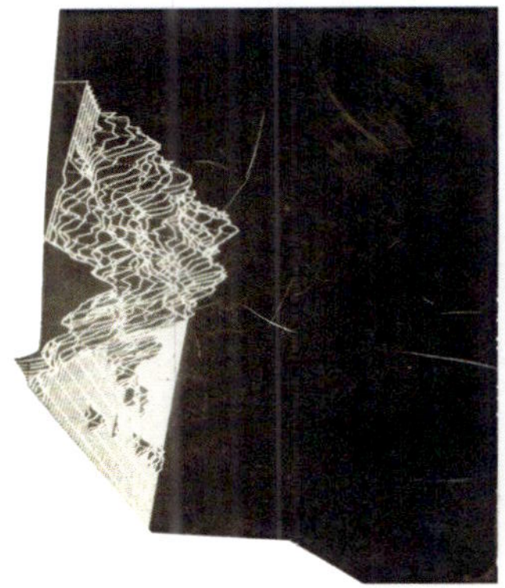

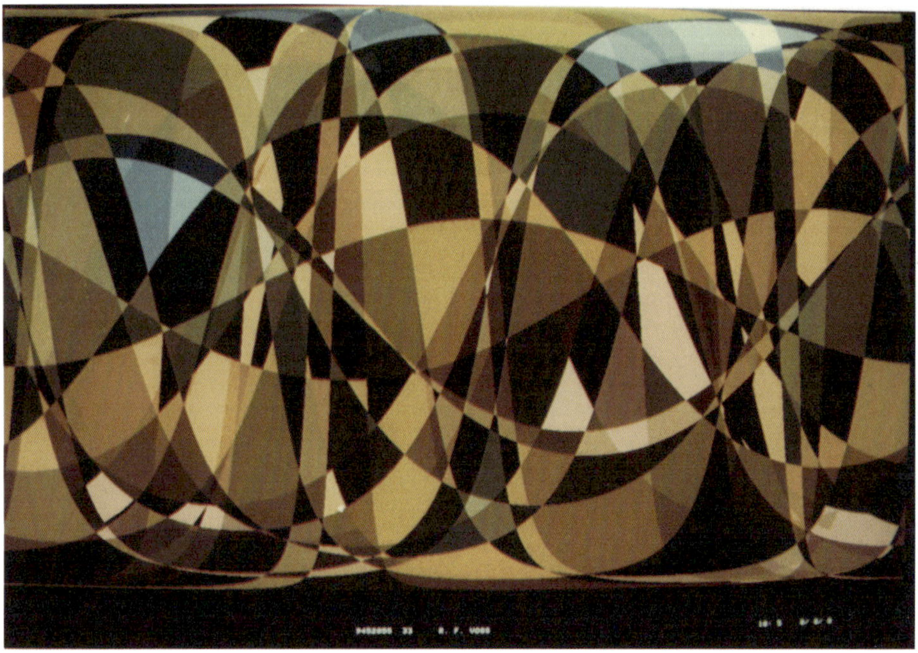

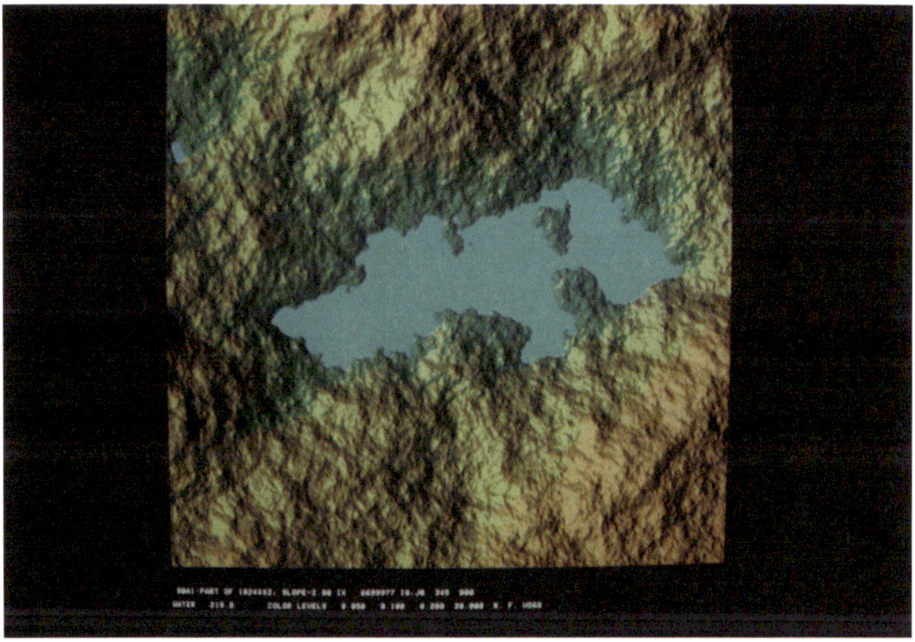

Cats. 9–12. Benoît Mandelbrot and Richard Voss. "Landscapes," August 1981. Color photographs. Collection Aliette Mandelbrot.

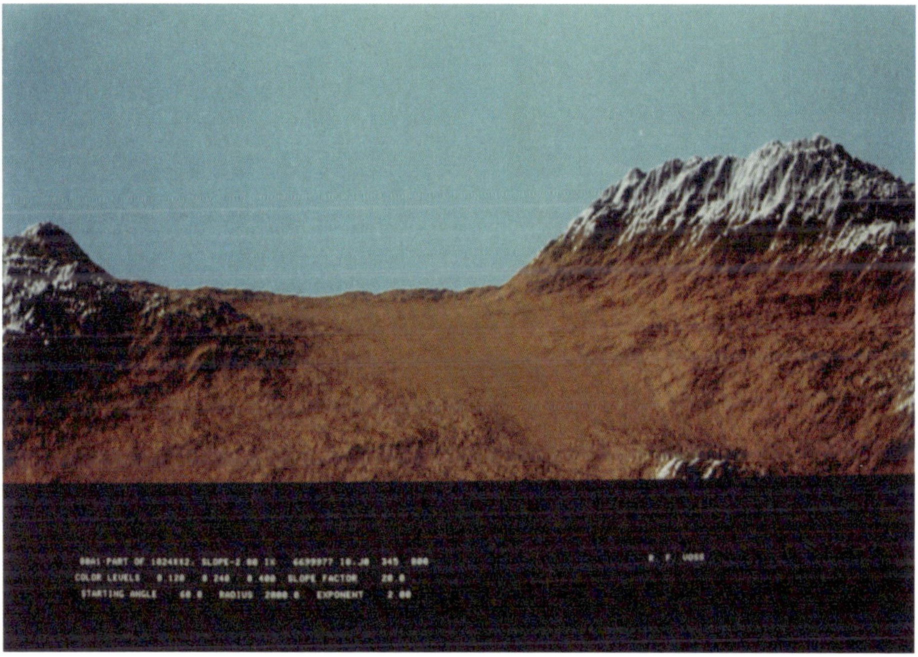
R. F. VOSS

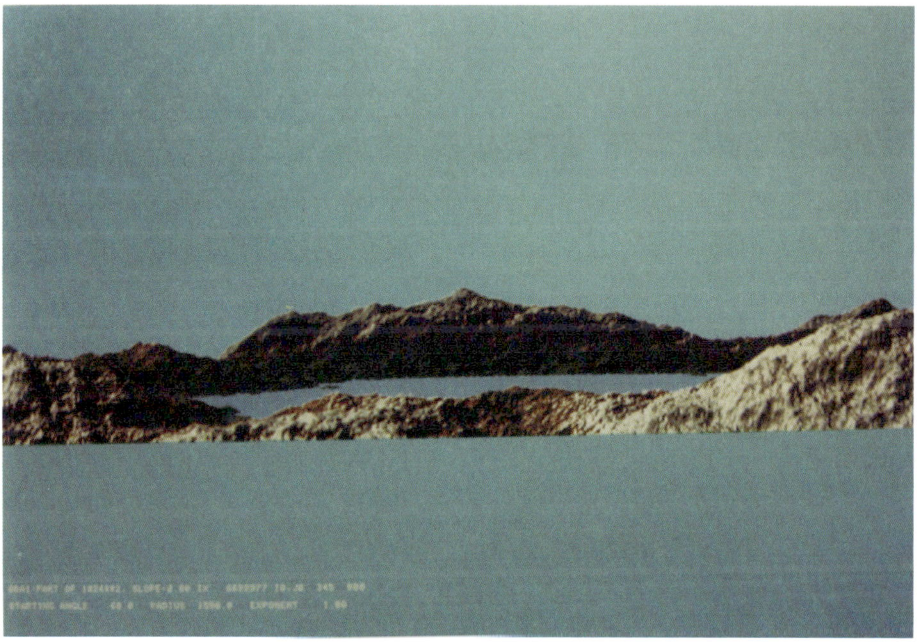

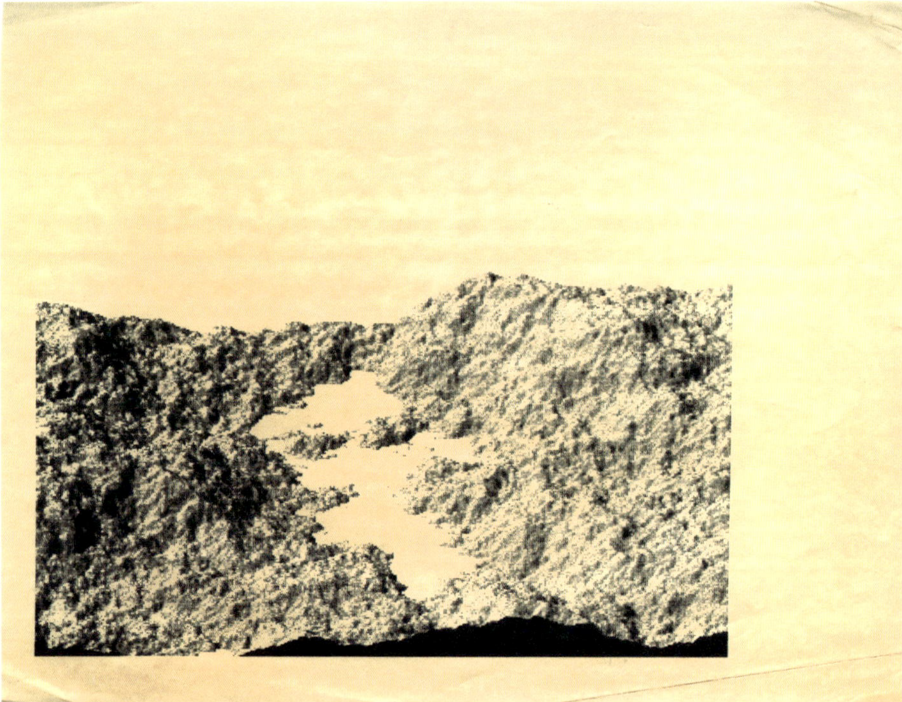

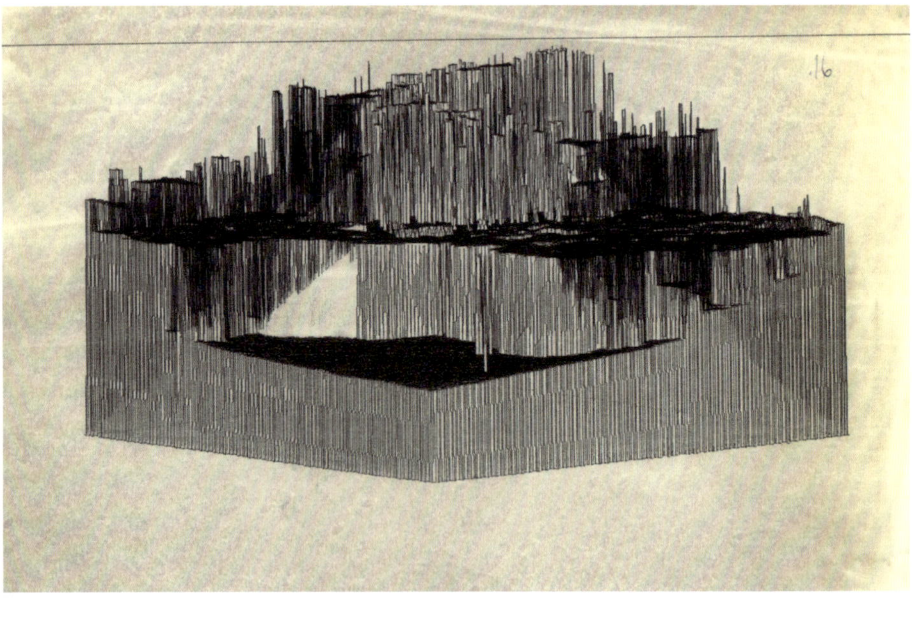

Cat. 13. (top) Benoît Mandelbrot and Richard Voss. "Landscape," ca. 1977. Computer-generated print. Collection Aliette Mandelbrot.

Cat. 14. (bottom) Benoît Mandelbrot and Sigmund Handelman, programmer. "Failed Nature," ca. 1974–77. Plotted graphics. Collection Aliette Mandelbrot.

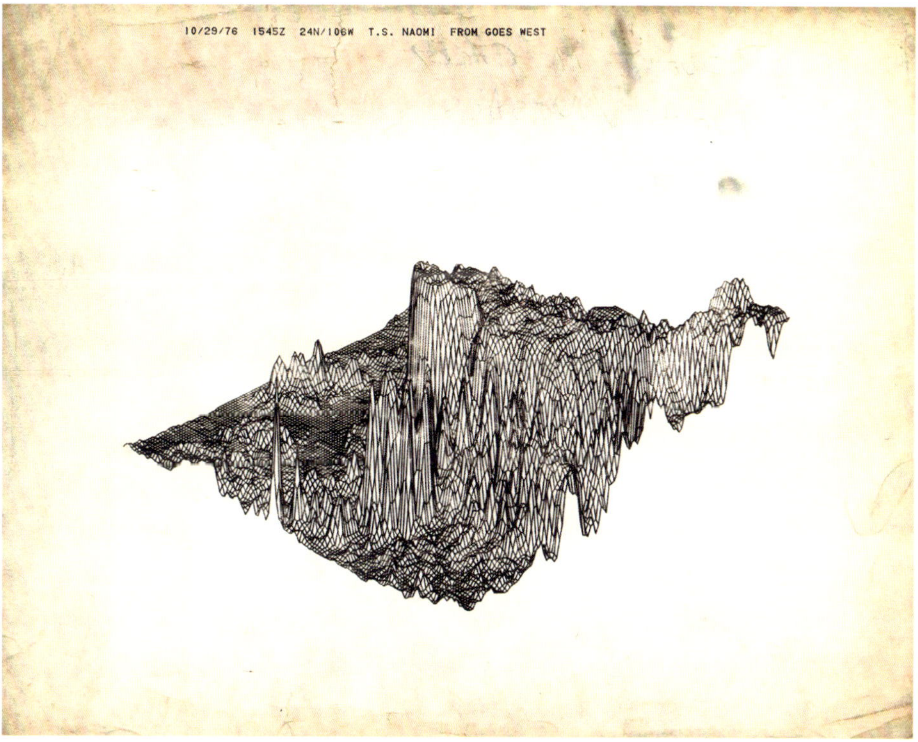

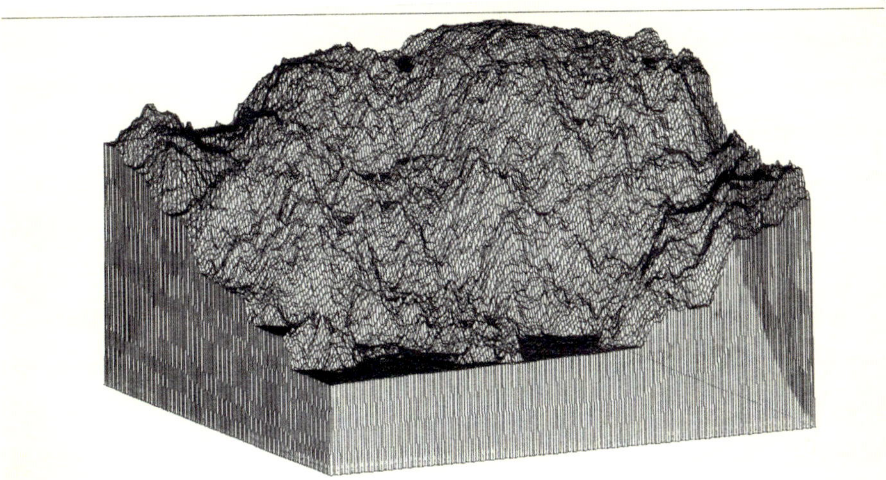

Cat. 15. (top) National Environmental Satellite, Data, and Information Service. “Meteorological Mountains,” October 29, 1976. Plotted graphics. Collection Aliette Mandelbrot.

Cat. 16. (bottom) Benoît Mandelbrot and Sigmund Handelman, programmer. “Fractal Mountain,” ca. 1974–77. Plotted graphics. Collection Aliette Mandelbrot.

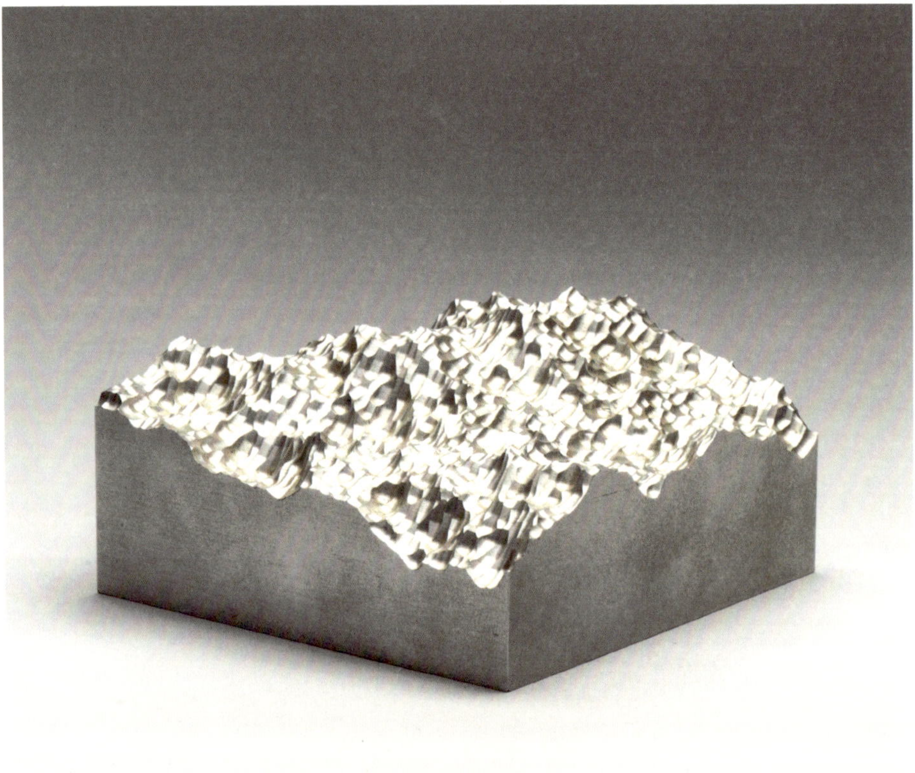

Cat. 17. Benoît Mandelbrot and Sigmund Handelman, programmer. “Carved Fractal Mountain,” ca. 1975–76. Metal object. Collection Aliette Mandelbrot.

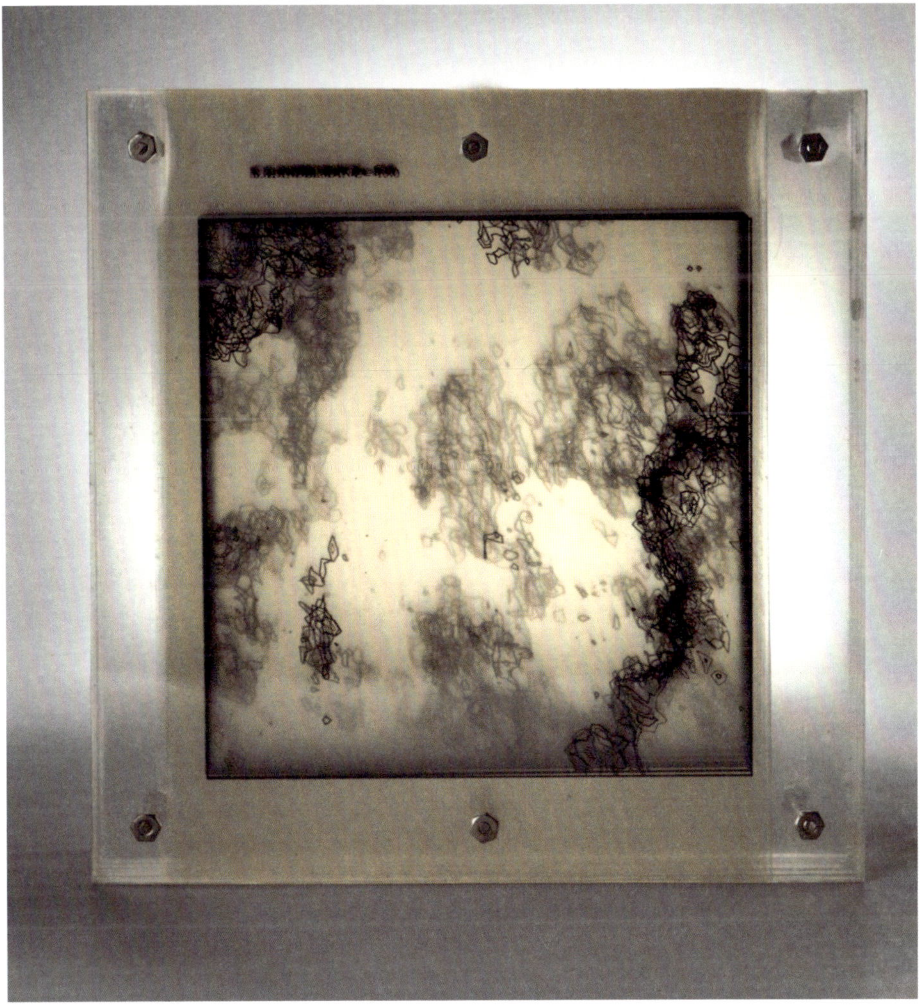

Cat. 18. "Diorama sculpture" based on computer-generated prints by Benoît Mandelbrot and Sigmund Handelman, programmer, poss. 1970s. Layered plexiglass block. Collection Aliette Mandelbrot.

The Fractal View: Nature in Mandelbrot's Geometry[1]

Jan von Brevern

I coined "fractal" from the Latin fractus, *which describes the appearance of a broken stone: irregular and fragmented.*[2]
—Benoît Mandelbrot

Nature is rough.[3]
—IBM

Fractal Nature

Is nature "rough"? Since antiquity, philosophers, theologians, and scientists have tried to demonstrate the regularity of nature. Mandelbrot's fractal geometry, by contrast, is based on the proposition that many natural phenomena are so irregular and fragmented that they cannot be described by means of Euclidean geometry.[4] Instead of approaching the forms of nature in the customary way—as regular, three-dimensional bodies—Mandelbrot made use of strange constructions that can no longer be accommodated within our familiar spatial dimensions: endlessly long curves that are nowhere smooth and in magnification constantly reveal new twists.[5] It was Mandelbrot's pathbreaking idea to call these mathematical oddities the fundamental model of nature.[6] A number of physicists, biologists, and geographers recognized the subversiveness in his proposal immediately—and enthusiastically accepted it. At last there was a way to describe in mathematical terms such varied phenomena as coastlines, vascular systems, clouds, and star clusters. It seemed that by way of fractal geometry, any number of problems could be scientifically investigated for the first time.[7] "Without fractal geometry, there would be no language to describe the results, except to say that they look like a mess."[8]

Apparently, this new "language" was needed if we were to obtain a new view of many natural phenomena.[9] Whereas previously, *order* had been sought, now it was discovered that there existed "disorder on all fronts."[10] Indeed, the availability of fractal geometry appears to have had enormous effects on the way we look at nature. In what follows, I mean to show that a whole new concept of nature was shaped by Mandelbrot's theory.[11]

Places Often Imagined Before

We have been oversold on nature's mathematical regularity.[12]
—Stephen Jay Gould

Just what was the new concept of nature introduced by Mandelbrot, and how did it differ from earlier concepts? These questions are illuminated by an advertisement produced by IBM in 1987 (fig. 1). It presents a computer graphic resembling a mountain landscape, with ridge lines, snow-covered peaks, and a lake in the valley below. The caption boasts: "This place is really nowhere." The accompanying text explains that the company gives its outstanding researchers considerable freedom because "the freedom to explore ideas can lead to places never imagined before." In connection with the mountain landscape, to be sure, these statements appear to overlook what makes Mandelbrot's graphics special. For what is new and exciting about them is hardly that they picture places that are "really nowhere"—one could say the same thing about any given landscape by Claude Lorrain (1600–1682). For centu-

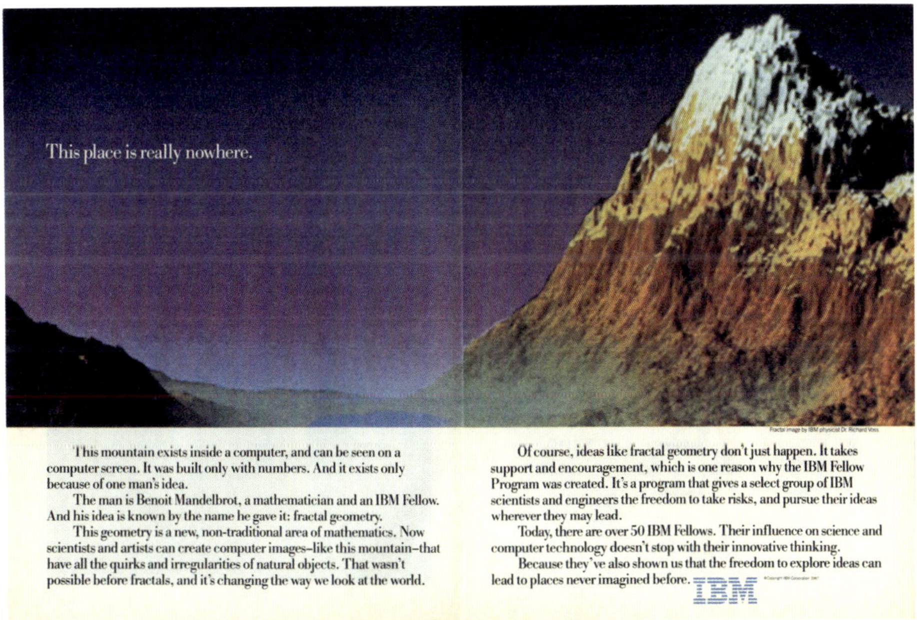

Fig. 1. "This place is really nowhere," 1987. Advertisement for IBM featuring a fractal landscape image created by IBM physicist Dr. Richard Voss. Reprint Courtesy of International Business Machines Corporation, © 2012 International Business Machines Corporation.

ries, it was taken for granted that landscape pictures had no corollary in reality: They were inventions that might rely on nature studies in their details but in their overall composition were subject solely to the laws of *vraisemblance*.[13] The second, metaphorical claim that this vista was "never imagined before" seems forced as well: One ordinarily imagines an ideal mountain landscape in virtually this same way.

Instead, what is actually expressed in these statements is astonishment at the birth of a new kind of picture, one that had not existed before, that looked amazingly like nature but was entirely created by a computer. It had to seem like a minor miracle that with the proper programming, totally abstract fractal algorithms could create an image of mountains and valleys like those seen in the Alps or the Rockies.[14] "Constantly," Mandelbrot marveled in reference to another, equally transporting fractal landscape, "I lapse into wondering during which trip I actually saw the vista. . . ."[15] But what does this verisimilitude mean? Could it possibly indicate an inner relationship between fractal geometry and nature? This is the kind of attractive but at the same time highly problematical conjecture that gives fractal geometry its allure.

In fact, "nature" is perhaps the most important term in Mandelbrot's theory; for him, it is nature that allegedly obeys a fractal logic and produces fractals. Mandelbrot's credo was that "there is a fractal face to the geometry of nature." He has quite explicitly criticized existing theories of geometry that have "turned away from nature."[16] Just how a landscape like the one in the IBM advertisement was created is shifted somewhat into the background. Not every fractal algorithm ultimately produced mountains and valleys. A long process of experimentation with various functions, the tiniest variations on the so-called "Hausdorff Dimension" D and various graphic options, produced a huge number of pictures of which only a very few ultimately resembled landscapes (see Found Images III, cats 1–8).[17] For that reason, there were fundamental doubts about Mandelbrot's theory early on. Mathematicians argued that his graphics had little to do with nature—that the only thing one could see and study in them were the (admittedly highly complex and very beautiful) intrinsic geometric shapes of algorithms.[18] There was also harsh criticism from art historians: To assume an essential structural similarity based on visual resemblance was seen to be an argument by analogy of the "crudest sort."[19]

Nevertheless, the suspicion prevailed that the sciences were being offered enormous opportunities. And quite apart from the issue of whether nature in fact obeys a fractal logic, one thing can be stated with certainty: Fractal geometry has effected a paradigm shift in our concept of nature.[20] For example, in the IBM ad from 1987 it was already assumed that natural objects are irregular: "Now scientists and artists can create computer images—like this mountain—that have all the quirks and irregularities of natural objects." This was a rejection of the postulate that nature is essentially orderly and always produces predictable forms, which had stood unchallenged since the beginning of the modern era.[21] As the philosopher Alfred North Whitehead (1861–1947) wrote in 1926, without the widespread conviction that there is an order to nature, there could have been no science.[22] But with the advent of fractal geometry—together with that of contemporary chaos theory, which during the 1980s was becoming increasingly prevalent[23]—scientists began to see a *disorderly* nature everywhere they looked. Reality itself had become irregular, and with the awareness of fractal geometry our view of it had, as it were, lost its innocence: "Only recently," it was noted in 1997, "there was no word to describe fractals. Today we are beginning to see such features everywhere. Tomorrow, we may look at the entire universe through a fractal lens."[24]

Landscapes were now being seen through a fractal lens as well. Suddenly, it seemed that one had been searching for geomorphological regularities for far too long. A new view was required, one that took into account the irregular aspects of land forms as well. Geoscientists were now convinced that "certain phenomena refuse to fit into a regular perspective."[25] Comparison with another IBM advertisement from a quarter-century earlier, in which the computer company seems to celebrate its mastery of the Euclidean paradigm, sheds light on the major changes in our thinking and observation that have taken place since then (fig. 2). The clear volumes seen in this 1960s ad—on which one can easily measure angles, place diagonals, and designate meridians—represent the basic building blocks of the scientific imagination. But the word "nature" is nowhere in evidence—here it seems that with logic alone, however brilliant, it is impossible to construct mountains.

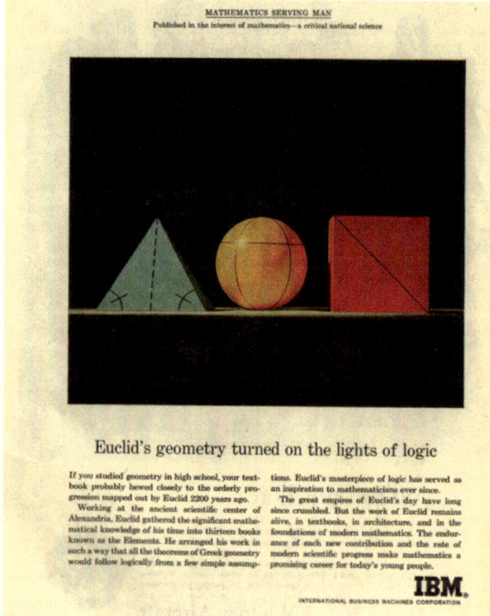

Fig. 2. "Euclid's geometry turned on the lights of logic," 1960. Advertisement for IBM. Reprint Courtesy of International Business Machines Corporation, © 2012 International Business Machines Corporation.

Ill-Figur'd Mountains

The nineteenth-century mathematicians may have been lacking in imagination, but nature was not.[26]
—Freeman Dyson

Mountains have always been difficult to fit into the order of nature. Using them as examples, it is possible to show how greatly our views of nature have changed over time—and also how those views have been dictated by prevailing geometry.

Early on, it was noted that mountains appear to be irregular, even chaotic. After a journey through the Alps, the English theologian Thomas Burnet (1635–1715) could assert that "there is nothing in Nature more shapeless and ill-figur'd than an old Rock or a Mountain, and all that variety that is among them, is but the various modes of irregularity."[27] In his *Telluris theoria sacra*, first published in 1681 and repeatedly reprinted up into the nineteenth century, Burnet tried to dispense with the prevailing doctrine that the earth was regularly formed. Mountains served him as a prime example, for "they have neither form nor beauty, nor shape, nor order, no more than the Clouds in the Air." In Burnet, one of the fundamental problems with disorder becomes apparent: Disorder is difficult to describe. Irregularity is defined as the absence of order and pattern, but there were no positive terms with which Burnet might characterize it. This was still the case even in the eighteenth century: The terminology available to Geneva's natural philosopher and Alpine explorer Horace-Bénédict de Saussure (1740–99), for example, was equally limited.[28]

Beginning in the nineteenth century, scholars were no longer willing to accept the irregularity of mountains. Geographers and geologists searched for regular structures behind their disorder, which they took to be only superficial. Such profound skepticism regarding the world of our direct perception was surely in part owing to Kant's distinction between the "thing in itself" and "things as they appear."[29] The German geographer Carl Ritter insistently warned against "taking the impression that strikes our senses for the object that produces it." One was by no means supposed to infer from such apparent confusion and lawlessness the actual absence of coherence and order. According to Ritter, what was needed was a "more profound consideration" of phenomena. Behind the planet's seemingly chaotic exterior there were doubtless higher symmetries and harmonies.[30]

The language available to scientists for the discovery of such symmetries and harmonies was geometry. Already, Ritter argued, the perfectly geometrical forms of cobwebs, plant cells, and crystals had been seen under the microscope. Geologists now went out in search of such regular patterns on a larger scale. Léonce Élie de Beaumont (1798–1874), one of the most influential figures in French science, postulated huge mountain systems that spread a pentagonal network—the so-called *réseau pentagonal*—across the entire globe (fig. 3). From the relative position of these mountain chains he hoped to be able to determine their age and reconstruct the conditions that had given rise to them.[31]

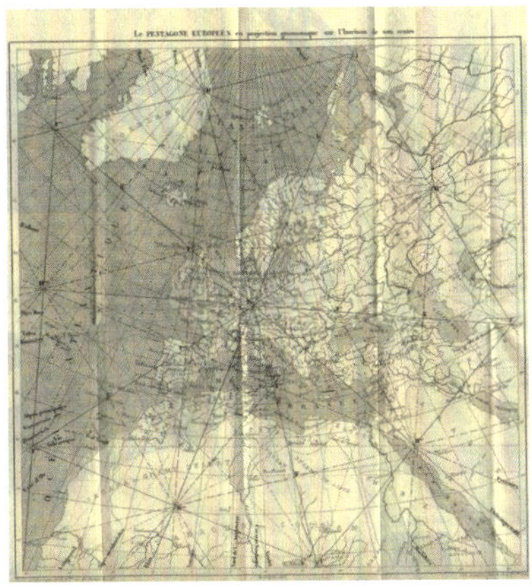

Fig. 3. Léonce Élie de Beaumont. "Le pentagone européen," 1852. Bibliothèque de l'École des mines de Paris, MINES ParisTech.

The belief that nature obeys geometrical principles runs through the entire nineteenth century. Perhaps it was most pointedly expressed by the French architect Eugène Viollet-le-Duc (1814–79), who engaged in geology for many years. While studying Mont Blanc, he noted that it was highly fortunate that geometry was invented before the worlds were created, "for without it, building these worlds would have been impossible" (fig. 4).[32]

Needless to say, such views decisively influenced the contemporary view of nature. And even though today such theories as Élie de Beaumont's *réseau pentagonal*

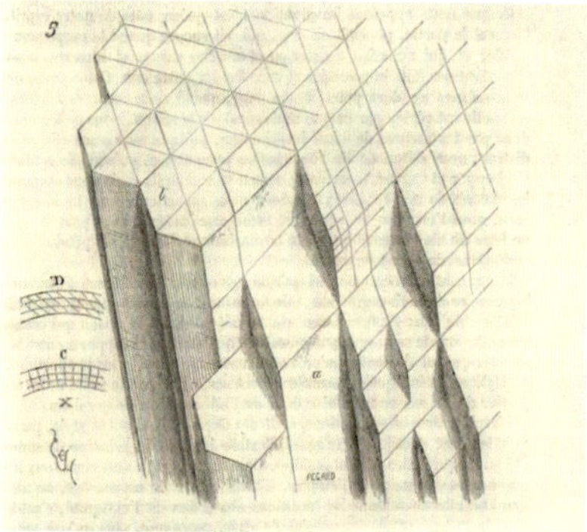

Fig. 4. Eugène Viollet-le-Duc. "Disposition des rhomboèdres granitiques." From Eugène Viollet-le-Duc, *Dictionnaire raisoné de l'architecture*, vol. 8 (Paris 1868), 481.

may strike us as absurd, they laid the groundwork for the tectonic theories of the twentieth century.[33] Plate tectonics, which finally came to be an acceptable geological theory in the 1960s, is still based on vast global structures that are only indirectly deduced from the planet's outward appearance. Although there was no longer any hope of finding perfect symmetries, to say nothing of pentagons, it was still possible to believe that one might find clearly defined forms that were compatible with Euclidean geometry.

Pattern Without Regularity

It is messy, and in the right way.[34]
—Michael Marder

One can now imagine what made fractal geometry seem like such an enormous provocation. In a sense, Mandelbrot was returning to the pre-Kantian theories of the seventeenth and eighteenth centuries. He ignored the whole distinction between "things as appearances" and "things in themselves," which had been a basic premise of modern scientific practice. Instead, like Burnet, he took the irregularities of natural phenomena at face value: Nature not only appears to be "irregular and fragmented," it *is* so. But unlike Burnet, he did not grow speechless in the face of disorder. On the contrary, with fractal geometry he developed a language for describing formerly indescribable phenomena. It is only his "morphology of the amorphous" that makes it possible for science to deal with disorder.[35]

To be sure, the irregularity of nature that Mandelbrot postulates can hardly be compared to that of the seventeenth and eighteenth centuries. It cannot be simply chance or "noise," otherwise it could not be captured in algorithms. Mandelbrot's achievement consists in having given a structure—a pattern—to fractal nature. This pattern is called "self-similarity." One recognizes fractals not in the forms themselves but in the invariance of their forms in scale: It does not matter how closely one zooms into the graphics, the same elements appear again and again.[36] Up close, it is claimed, coastlines would look precisely as irregular and random as from a great distance. Self-similarity seems like a revival of the notion of the *macrocosmos in microcosmo* so beloved by the ancients and by the hermeticists of early modern times, the idea that the whole cosmos is mirrored in the tiniest of objects and organisms.[37] But identifying it as the inherent pattern of natural irregularity—that would wait for Mandelbrot's insight.

Scientists have had to become accustomed to this unfamiliar "idea of pattern without regularity," which is also essential to the theory of deterministic chaos.[38] By now it appears that the "fractal lens" can hardly be set aside: Today, nature unquestionably appears to us as far "rougher" than it did half a century ago. It almost seems as if Mandelbrot had not only invented a new geometry—but a new nature as well.

1 I could not have written this text if Nina Samuel had not permitted me to read her dissertation on Mandelbrot before publication: many thanks to her. I also thank Katja Müller-Helle for her critical comments and helpful suggestions.

2 Benoît Mandelbrot, *The Fractal Geometry of Nature* (New York: Freeman, 1982), 262.

3 http://www.ibm.com/ibm100/us/en/icons/fractal/ (accessed March 28, 2012).

4 Benoît Mandelbrot, *Fractals: Form, Chance, and Dimension* (San Francisco: Freeman, 1977), 1.

5 Mathematicians here speak of functions that are constant but nowhere differentiable.
6 See Freeman Dyson, "Characterizing Irregularity: Review of Fractals. *Form, Chance and Dimension* by Benoît B. Mandelbrot," *Science* 200 (1978). Mathematicians frequently described functions like the Peano Curve as being "pathological"; see Mandelbrot, *The Fractal Geometry of Nature*, 59.
7 "It's only with Mandelbrot's insight . . . that physicists were able eventually to start to attack those problems." Michael F. Shlesinger, cited in Ivars Peterson, "Fractal Past, Fractal Future," *Science News*, no. 9 (1997): S13.
8 Michael Marder, "Roughing It," *Science* 277 (1997): 647.
9 Fractal geometry has often been described as a language; see, for example, Heinz-Otto Peitgen, Hartmut Jürgens, and Dietmar Saupe, *Chaos and Fractals: New Frontiers of Science* (New York: Springer, 1989), 216.
10 Mark Buchanan, "The Irregularity of Reality," *Nature Physics* 7 (2011): 184.
11 For changes in the concept of nature throughout history, see Dieter Groh and Ruth Groh, *Weltbild und Naturaneignung. Zur Kulturgeschichte der Natur* (Frankfurt am Main: Suhrkamp, 1991); also, Götz Grossklaus and Ernst Oldemeyer, eds., *Natur als Gegenwelt. Beiträge zur Kulturgeschichte der Natur* (Karlsruhe: Loeper, 1983).
12 Stephen Jay Gould, *Questioning the Millennium: A Rationalist's Guide to a Precisely Arbitrary Countdown* (London: Vintage, 1998), 120.
13 See, for example, Oskar Bätschmann, *Entfernung der Natur. Landschaftsmalerei 1750–1920* (Cologne: DuMont, 1989). The Netherlandish tradition of landscape painting is an exception; see the exemplary discussion in Max Imdahl, "Jacob van Ruisdaels 'Die Mühle von Wijk,'" in *Max Imdahl: Gesammelte Schriften*, ed. Gundolf Winter (Frankfurt am Main: Suhrkamp, 1996).
14 See Nina Samuel, "'I Look, Look, Look, and Play with Many Pictures,' Zur Bilderfrage in Benoît Mandelbrot's Werk," in *Verwandte Bilder. Die Fragen der Bildwissenschaft*, ed. Ingeborg Reichle, Steffen Siegel, and Achim Spelten (Berlin: Kadmos, 2007).
15 Mandelbrot, *The Fractal Geometry of Nature*, 271.
16 Ibid., 1, 3.
17 Among other ways, fractals are distinguished by the fact that they have a non-integral dimension D (in contrast to such traditional geometrical figures as line, surface, or solid, which always have integral dimensions). For his mountains, Mandelbrot varied the dimension between D=2.1 and D=2.5; ibid. 256–71.
18 See James W. Cannon, "Review of *The Fractal Geometry of Nature* by Benoît B. Mandelbrot," *The American Mathematical Monthly* 91 (1984): 594–98.
19 Horst Bredekamp, "Mimesis, grundlos," *Kunstforum International* 114 (1991): 278–88.
20 For the concept of the "paradigm shift," see Thomas S. Kuhn, *The Structure of Scientific Revolutions* (Chicago: University of Chicago Press, 1962).
21 See Lorraine Daston and Michael Stolleis, eds., *Natural Law and Laws of Nature in Early Modern Europe: Jurisprudence, Theology, Moral, and Natural Philosophy* (Farnham: Ashgate, 2008), and Michael Hampe, *Eine kleine Geschichte des Naturgesetzbegriffs* (Frankfurt am Main: Suhrkamp, 2007).
22 Alfred North Whitehead, *Science and the Modern World* (Cambridge: Cambridge University Press, 1926), 4.
23 See James Gleick, *Chaos: Making a New Science* (New York: Viking, 1987).
24 Peterson, "Fractal Past, Fractal Future," S13.
25 W. E. H. Culling, "A New View of the Landscape," *Transactions of the Institute of British Geographers* 13 (1988): 346.
26 Dyson, "Characterizing Irregularity," 678.
27 Thomas Burnet, *The Sacred Theory of the Earth, Containing an Account of the Original of the Earth* (London, 1681), 98.
28 See, for example, Horace-Bénédict de Saussure, *Voyages dans les Alpes, précédés d'un essai sur l'histoire naturelle des environs de Genève* (Neuchâtel: Samuel Fauche, 1796),179. For Saussure, see also Claude Reichler, *La découverte des Alpes et la question du paysage* (Geneva: Georg Editeur, 2002).
29 Immanuel Kant, *Kritik der reinen Vernunft* (Riga: Hartknoch, 1787), xxv. For the influence of Kant's distinction on scientific observation, see also Jonathan Crary, *Techniques of the Observer: On Vision and Modernity in the Nineteenth Century* (Cambridge, MA: MIT Press, 1990), 69.
30 Carl Ritter, "Ueber räumliche Anordnungen auf der Aussenseite des Erdballs und ihre Functionen im Entwicklungsgange der Geschichten," in *Einleitung zur allgemeinen vergleichenden Geographie*, ed. Carl Ritter (Berlin: Georg Reimer, 1852), 209–10.
31 Léonce Élie de Beaumont, *Notice sur les systèmes des montagnes*, 3 vols. (Paris: P. Bertrand, 1852). See also Barbara A. Kennedy, *Inventing the Earth: Ideas on Landscape Development since 1740* (Malden, MA: Blackwell, 2006), 10; and Jaques Touret, "Élie de Beaumont (1798–1874), des systèmes de montagnes au réseau pentagonal," in *L'essor de la géologie française*, ed. Jean Gaudant (Paris: Presses des Mines, 2009).
32 Viollet-le-Duc, quoted from Pierra A. Frey and Lise Grenier, eds., *Viollet-le-Duc et la Montagne* (Grenoble: Glénat, 1993), 29. For his geological studies, see Eugène Emmanuel Viollet-le-Duc, *Le Massif du Mont-Blanc* (Paris: Baudry, 1876).
33 See Mott T. Greene, *Geology in the Nineteenth Century: Changing Views of a Changing World* (Ithaca, NY: Cornell University Press, 1982), 77.
34 Marder, "Roughing It," 647.
35 Mandelbrot, *The Fractal Geometry of Nature*, 1.
36 A demonstration video of a zoom into a Mandelbrot set can be found on the IBM Web site: http://www.ibm.com/ibm100/us/en/icons/fractal/.
37 See Robert Fludd, *The Origin and Structure of the Cosmos*, trans. Patricia Tahil (Edinburgh: Magnum Opus Hermetic Sourceworks, 1982). This was also a fundamental concept behind the creation of the modern-art and curio cabinet: See Andreas Grote, ed., *Macrocosmos in Microcosmo: Die Welt in der Stube. Zur Geschichte des Sammelns 1450 bis 1800* (Oplanden: Leske + Budrich, 1994).
38 Culling, "A New View of the Landscape," 357.

Found Images IV

Pictures of Chance

Nina Samuel

You cannot draw the sky at all; for the clouds will not fit if you draw one part of them three or four minutes before another.[1]
—John Ruskin

Chance is synonymous with the unpredictably changeable, the causally incomprehensible, the incalculable, the fluctuating: Clouds are ideal symbols of a chance absence of form in constant change. The philosopher Michel Serres came to the conclusion that the shapes of clouds could be traced only with difficulty or not at all: "No one can draw the edge of a cloud, this margin of chance. In which the tiny components quiver and dissolve, at least to our eyes. . . . Based on these wholly new boundaries, which are abandoned by drawing and geometry, a new world will soon enough discover the disintegration and diffusion of atoms and molecules."[2]

Yet it was the cloud, so inimical to geometry, that became the prototype of Mandelbrot's science (cats. 1–3). Etymologically, the word "fractal" not only suggests a broken or jagged structure, it also explicitly refers to a basic meteorological concept that takes into account both the movement of clouds and their morphology: "*Fractus*: A cloud species in which the cloud elements are irregular but generally small in size, and which presents a ragged, shredded appearance, as if torn. All of these characteristics change ceaselessly and often rapidly."[3] Fully cognizant of this definition, Mandelbrot also called his geometry a science of non-shapes—a "morphology of the amorphous"[4]—and in so doing once again set it apart from Euclid's. His assertion that "clouds are not spheres, mountains are not cones, coastlines are not circles" has been frequently quoted and paraphrased.[5] It suggests that one ought to devote oneself to the irregular shapes beyond the reach of the compass, inasmuch as they are "*grainy, hydralike, in between, pimply, pocky, ramified, seaweedy, strange, tangled, tortuous, wiggly, wispy, wrinkled.*"[6]

Mandelbrot's working materials revealed that he collected various kinds of cloud pictures in a special file, which included a private travel photograph of his wife's *California Summer* 1976, a stock-exchange advertisement with a smoking volcano cut out of a magazine, and a high-contrast, blue-and-white computer simulation that his co-worker Shaun Lovejoy had made in the early 1980s (circa 1984).[7] He related these various levels of reality to each other and compared them side by side. By creating analogies across disciplines, Mandelbrot suggested both that there was a universally valid principle behind all the various pictorial phenomena and that processes could be described mathematically. The universality of the line interrupted by chance, which also characterizes the margins of clouds, stood for the universality of his fractal theory.

Beginning in the 1960s, both artists and scientists were testing the computer as an image-making device. Graphic experiments with the laws of chance were undertaken early on.[8] The visit to Mandelbrot's office revealed an almost excessive interest in the pictorial possibilities of programming chance lines. His extensive series of pictures on this subject filled a number of boxes. In an example from 1978 (cat. 4), a line extends across the entire sheet in such a way that two focal points are created in the upper left and lower right. The line intersects itself again and again, forming nodal points that suggest the hectic path of a small insect. The composition is both rhythmic and erratic, and takes on a temporal dimension thanks to the alternation of long and short stretches. In the upper left are the initials (MRL) of the programmer, Mark R. Laff, who worked with Mandelbrot on this chance picture, as well as the name of the graphic program, LBLGRAPH. The rule by which chance became visualized could even bear subtle features of authorship: In this case, what generated the pseudo-chance, Mandelbrot's private telephone number in Scarsdale, New York, appears directly above his subsequent signature: "SEED = 4725916."[9]

Mandelbrot's "probability pictures" (cats. 4–31) were related to the increasing influence of probabilistic methods in the sciences, which researchers deployed to look for ways to give form to chance.[10] A prominent feature of this endeavor was a sudden new interest in Brownian molecular motion, also known in mathematical statistics since 1905 as the concept of the "random walk."[11] The works of the French mathematician Paul Lévy, a pupil of Henri Poincaré and one of Mandelbrot's most important teachers at Paris's École polytechnique, had made important contributions to this field in the 1940s.[12] Mandelbrot took that interest with him to his workplace at IBM: In computer-generated form, Brownian motion became one of the phenomena on which his programming of chance pictures was based. In homage to his teacher, he gave the name "Lévy flight" to a series of pictures that translated this mathematics of probability theory into pictures (cats. 4–7 and 12–25).[13]

Mandelbrot's computer simulations based on Brownian motion were central to his creation of connections between chance, visualized mathematics, and nature: Such "probable pictures" would become representations of nature. In these images, for example, Mandelbrot realized a cosmological model formulated by Fred Hoyle (cats. 26–31).[14] In order to appreciate his pictures as nature simulations, however, it was necessary to regularly study various views side by side. This is important, for without comparing pictures, it would not have been possible to document the connection between geometry and natural phenomena at all. There was actually no direct relationship with cosmic reality to be found there, but *thanks to the representations*, the validity of the basic theses is nevertheless immediately apparent. It was the picture that produced the reference.

1 John Ruskin, *The Elements of Drawing: In Three Letters to Beginners* (New York: Smith, Elder, & Co, 1857), 188.
2 Michael Serres, *Über Malerei. Vermeer–La Tour–Turner* (Dresden: Verlag der Kunst, 1995), 97.
3 Ralph E. Huschke, ed., *Glossary of Meteorology* (Boston: American Meteorological Society, 1959), 231.
4 Benoît Mandelbrot, *The Fractal Geometry of Nature* (New York: Freeman, 1982), 1. This is reminiscent of the "form of the formless" in Art Informel; see Gottfried Boehm, "The Form of the Formless: Abstract Expressionism and Art Informel," in *Action Painting*, ed. Delia Ciuha (Ostfildern: Hatje Cantz, 2008).
5 Mandelbrot, *The Fractal Geometry of Nature*, 1.
6 Ibid., 5 (italics Mandelbrot).
7 See Shaun Lovejoy and Benoît Mandelbrot, "Fractal Properties of Rain, and a Fractal Model," *Tellus Series A* 37A (May 1, 1985).
8 Barbara Nierhoff, "Die Ordnung des Zufalls. Ein Blick auf Prinzipien früher digitaler Kunst am Beispiel von Georg Nees," in *Georg Nees, Künstliche Kunst: Die Anfänge*, ed. Barbara Nierhoff (Bremen: Kunsthalle, 2005), 6–13; Barbara Nierhoff-Wielk, "Ex Machina–Die Begegnung von Computer und Kunst. Ein Blick zurück," in *Ex Machina–Frühe Computergrafik bis 1979* (Berlin: Deutscher Kunstverlag, 2007), 34–40; and Margit Rosen and Peter Weibel, "Die Zukunft eines künstlerischen Mediums," in *Ex Machina*, 197–99.
9 Richard Voss, e-mail to the author, June 1, 2011.
10 For the growing influence of probability representations and statistical methods in the sciences in the twentieth century, see Gerd Gigerenzer et al., *The Empire of Chance: How Probability Changed Science and Everyday Life* (Cambridge: Cambridge University Press, 1990); for physics, see pages 185–223.
11 See Karl Pearson, "The Problem of the Random Walk," *Nature* 72 (1905): 294.
12 Paul Lévy, *Processus stochastiques et mouvement brownien* (Paris: Gauthier-Villars, 1948); see also Benoît Mandelbrot, *Fractals: Form, Chance, and Dimension* (San Francisco: Freeman, 1977), 265–68. For his use of Lévy in economics, see Benoît Mandelbrot, "Sur certains prix spéculatifs: faits empiriques et modèle basé sur les processus stables additifs de Paul Lévy," *Comptes Rendus* (Paris) 254 (1962): 2968–3970.
13 See Mandelbrot, *Fractals*.
14 See Fred Hoyle, *Astronomy and Cosmology: A Modern Course* (San Francisco: W. H. Freeman & Co., 1975).

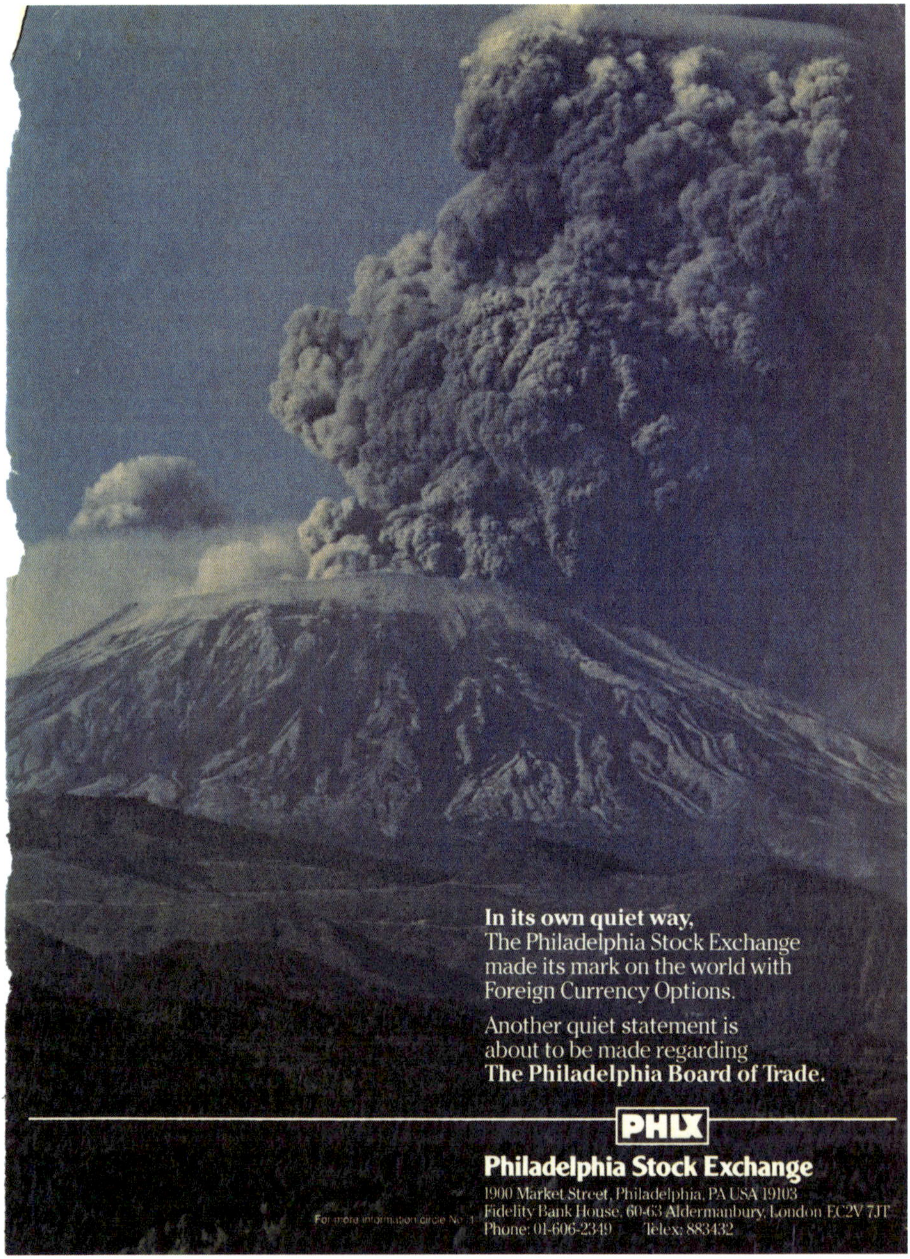

Cat. 1. "In its own quiet way." Advertisement for Philadelphia Stock Exchange. Unknown date. Collection Aliette Mandelbrot.

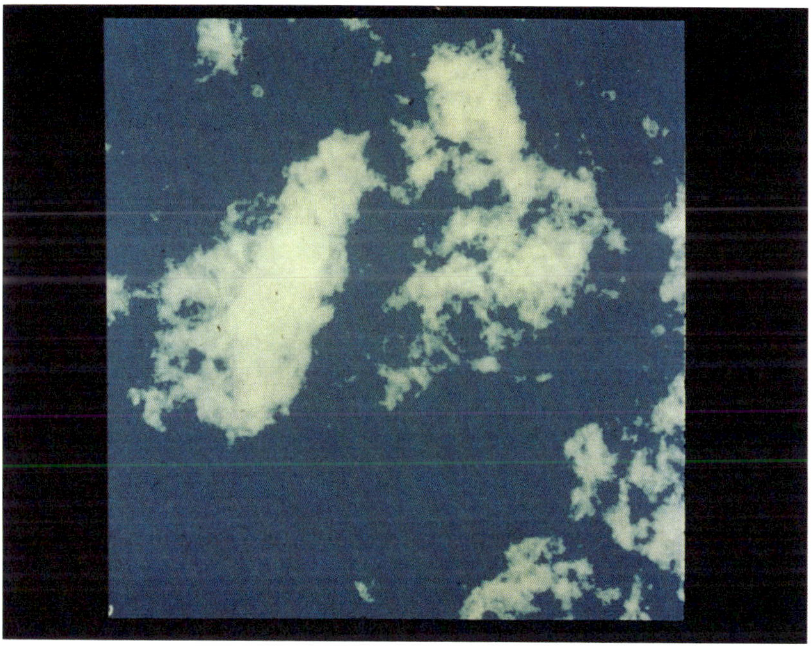

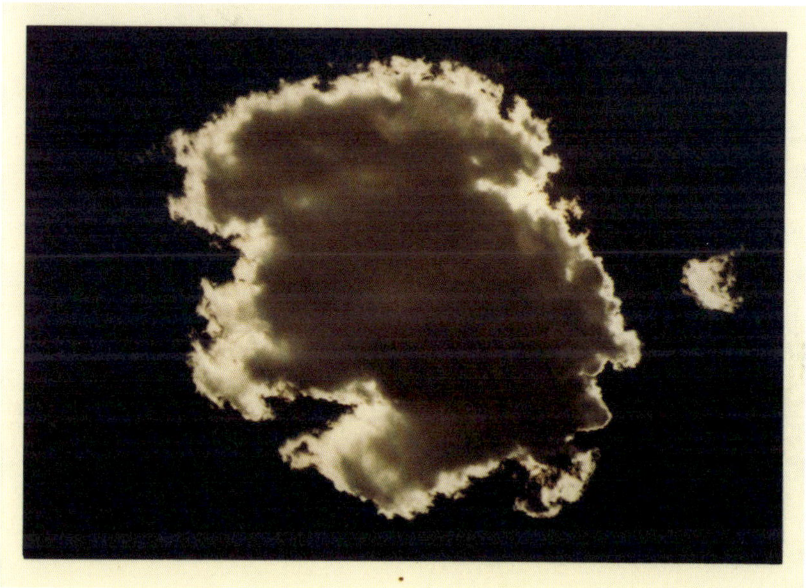

Cat. 2. (top) Shaun Lovejoy and Benoît Mandelbrot. Image from related article: "Fractal Properties of Rain and a Fractal Model," 1985. Photograph from computer-generated simulation. Collection Aliette Mandelbrot.

Cat. 3. (bottom) Aliette Mandelbrot. *California Summer*, 1976. Photograph. Collection Aliette Mandelbrot.

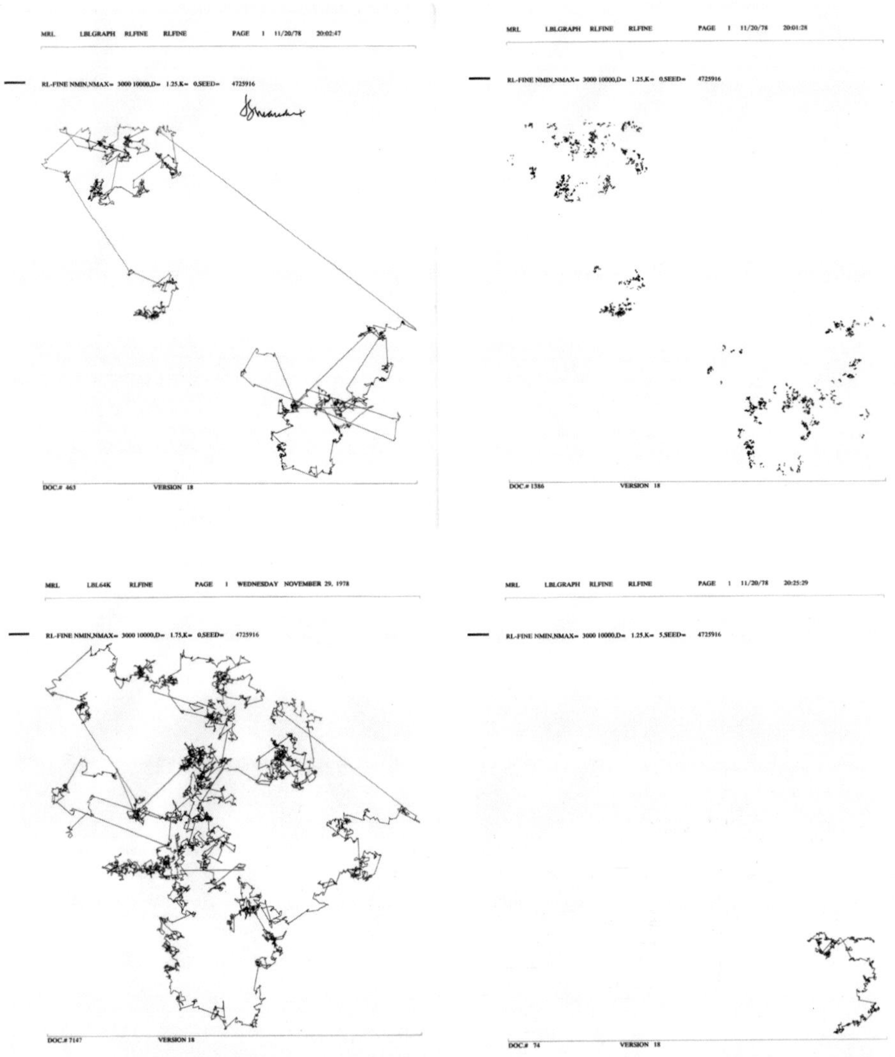

Cats. 4–7. Benoît Mandelbrot and Mark R. Laff, programmer. *Lévy Flight*, 1978. Computer-generated prints on photographic paper. Collection Aliette Mandelbrot.

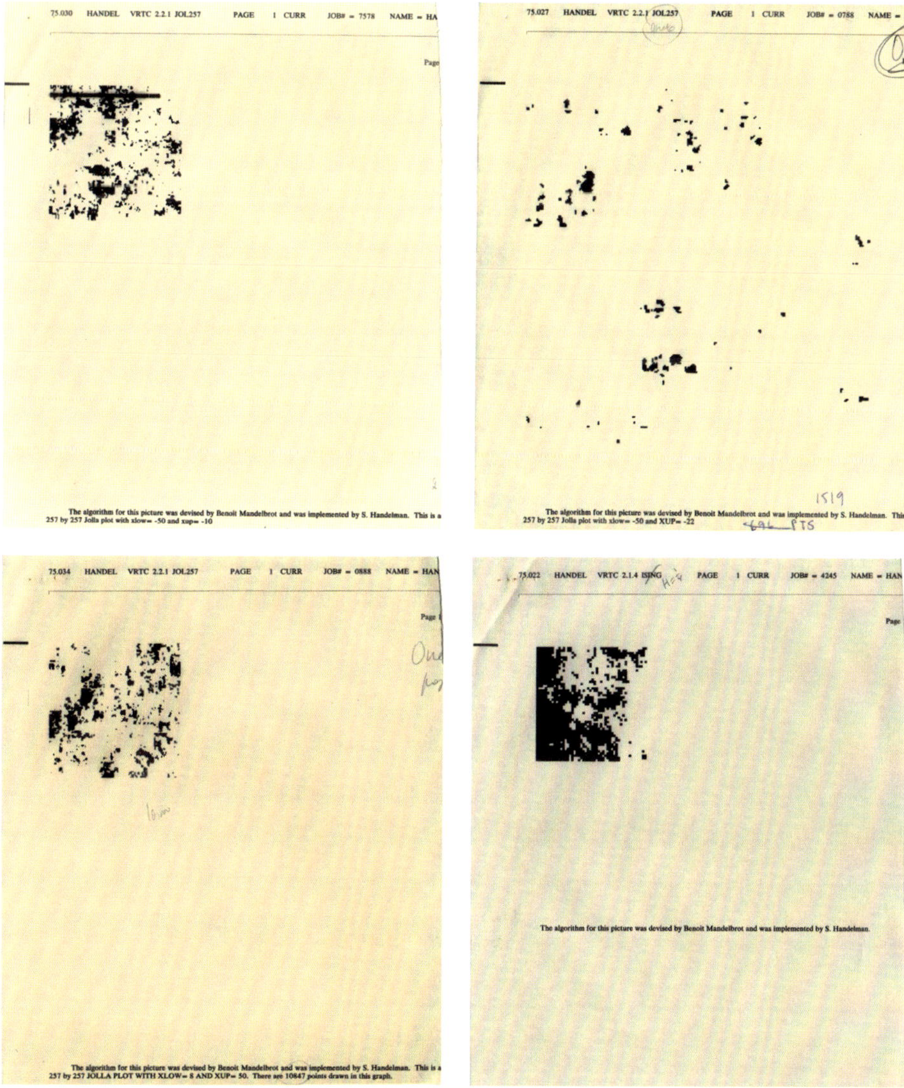

Cats. 8–11. Benoît Mandelbrot and Sigmund Handelman, programmer. "Random distributions," ca. 1974–77. Computer-generated prints on photographic paper. Collection Aliette Mandelbrot.

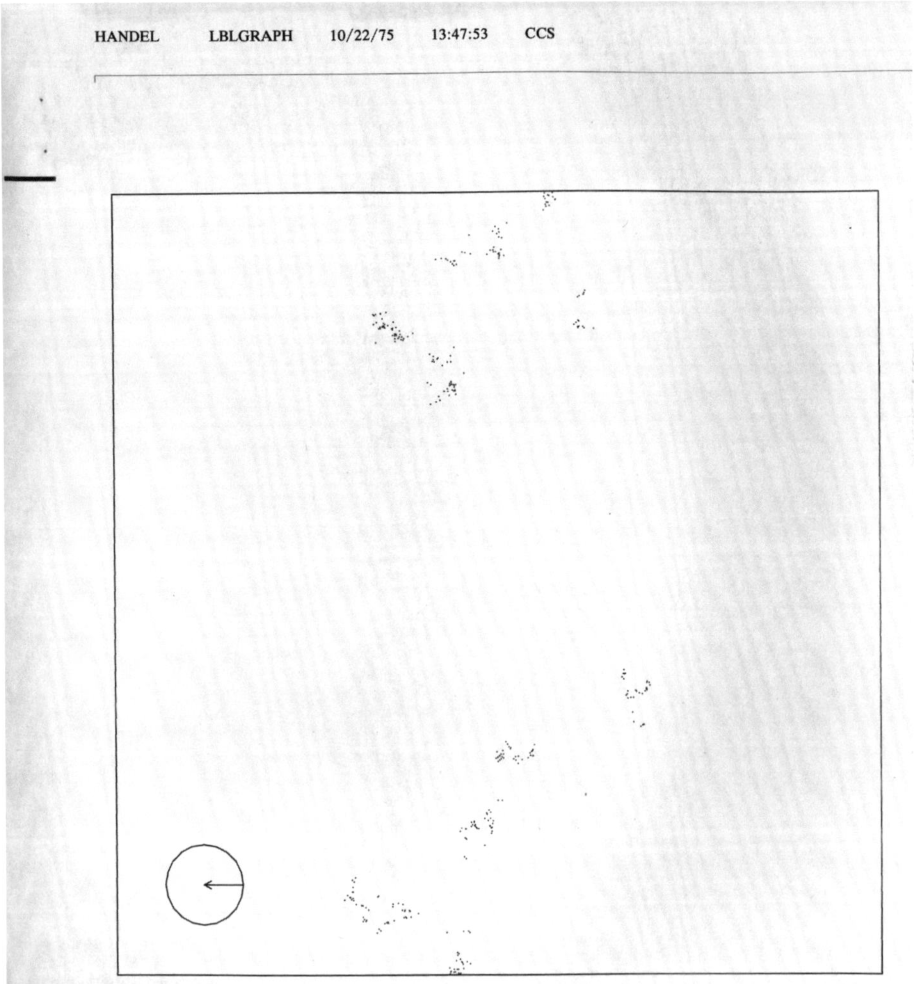

Cats. 12–13. Benoît Mandelbrot and Sigmund Handelman, programmer. *Lévy Flight*, 1975. Computer-generated prints on photographic paper. Collection Aliette Mandelbrot.

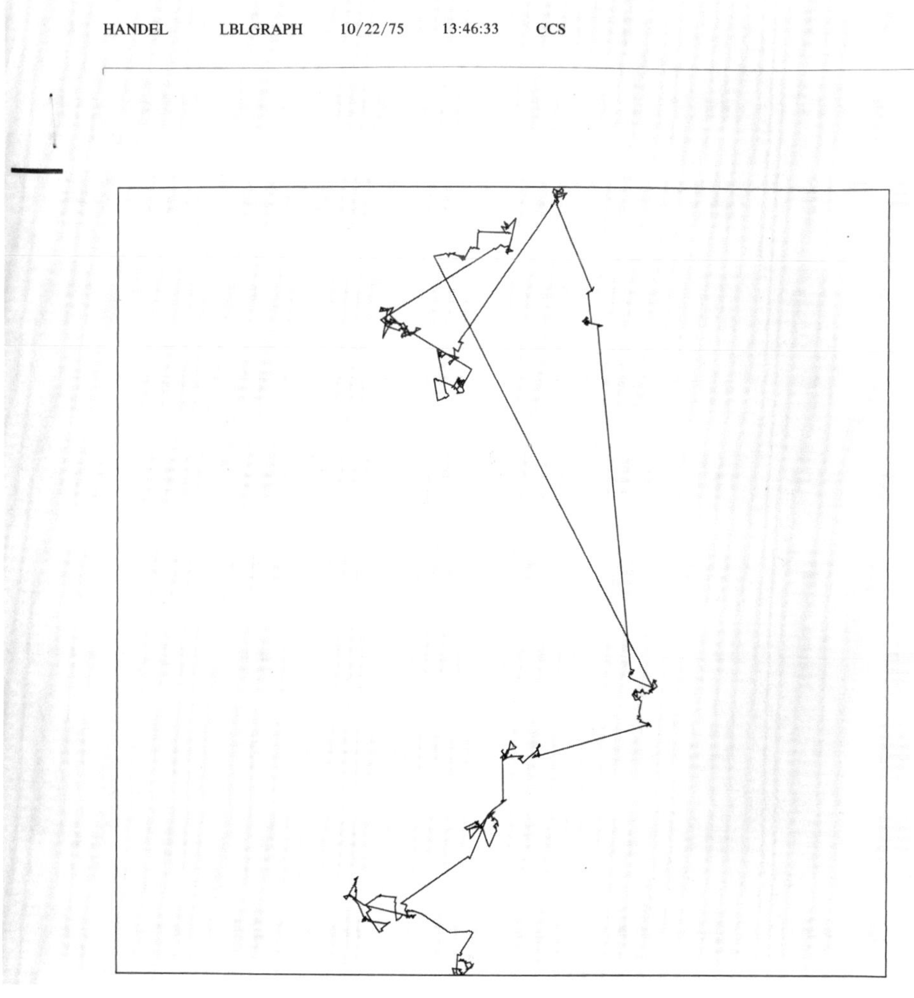
HANDEL LBLGRAPH 10/22/75 13:46:33 CCS

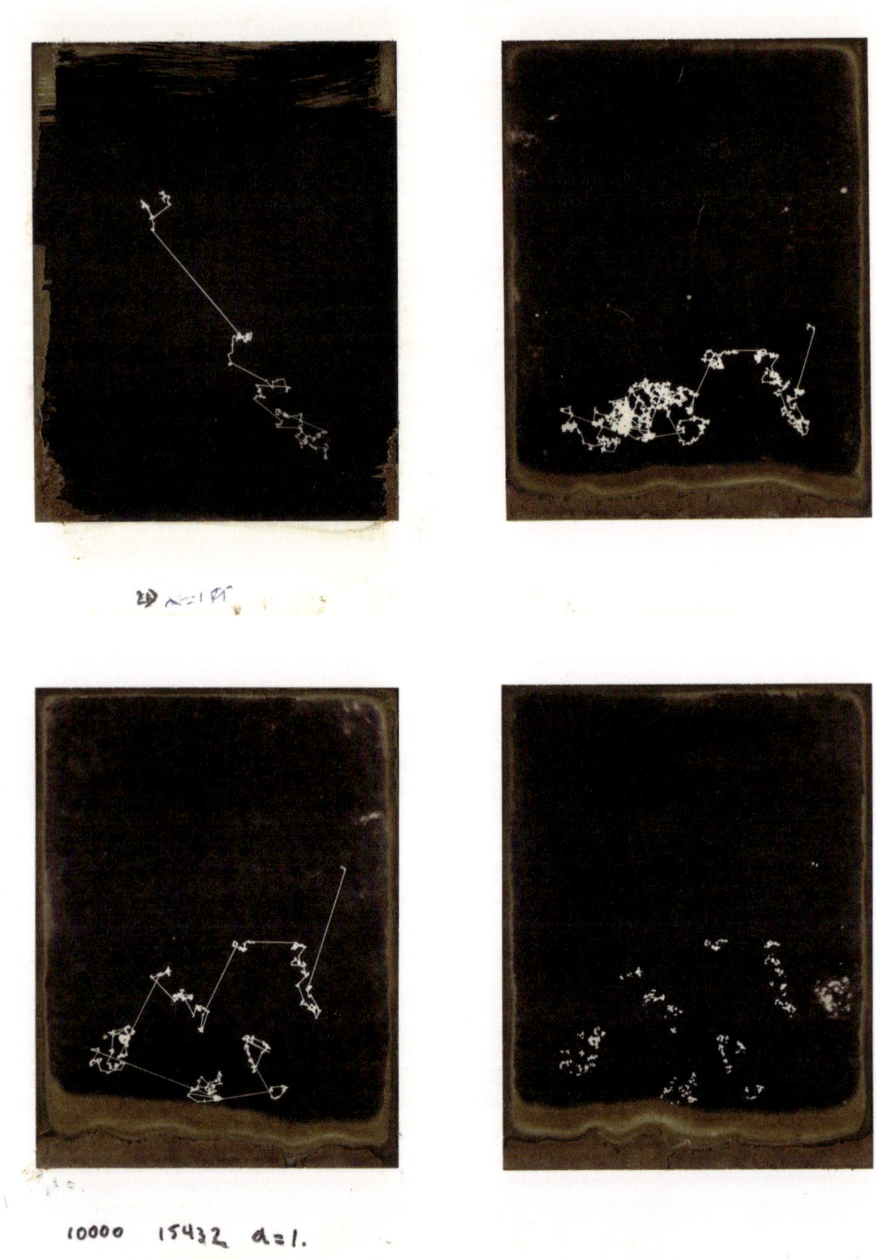

Cats. 14–25. Benoît Mandelbrot and Sigmund Handelman, programmer. *Landscapes and Galaxies*, series of forty-five, 1974–77. Polaroid photographs. Collection Aliette Mandelbrot.

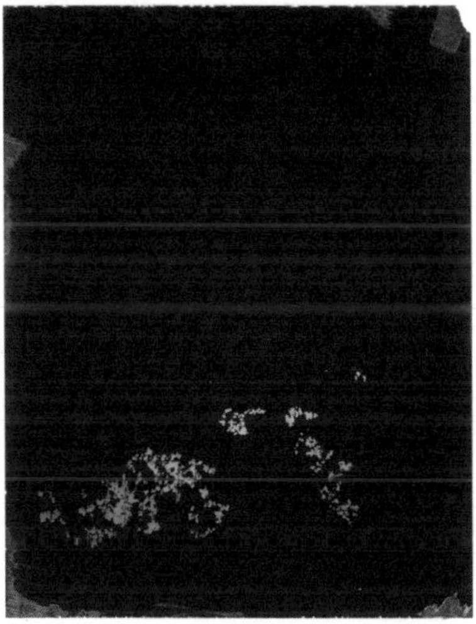

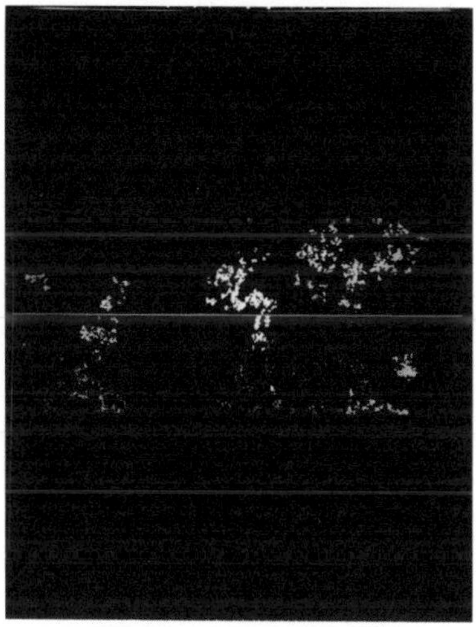

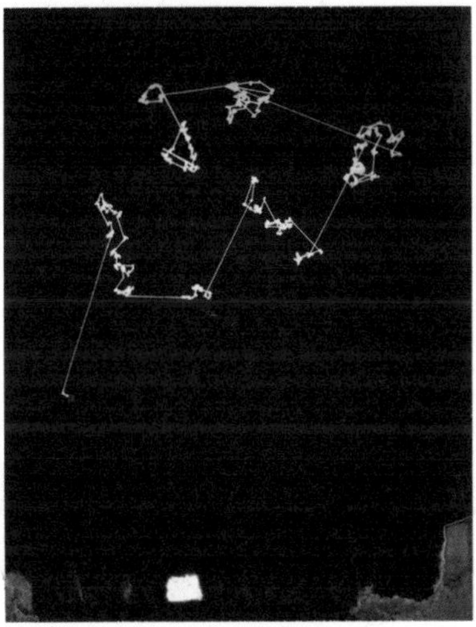

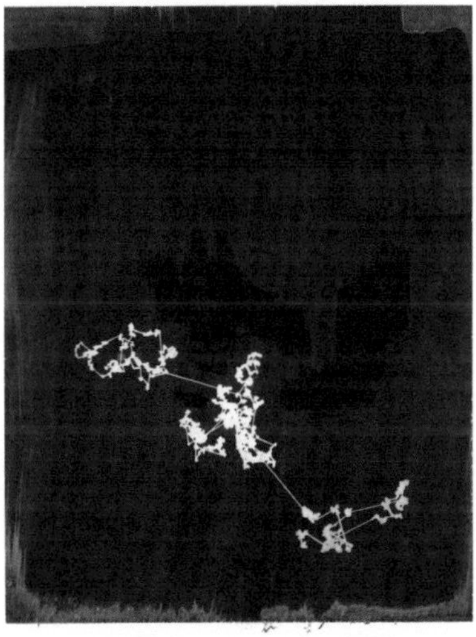

15432 2D 10000 α=1

87181 α=1.5
20

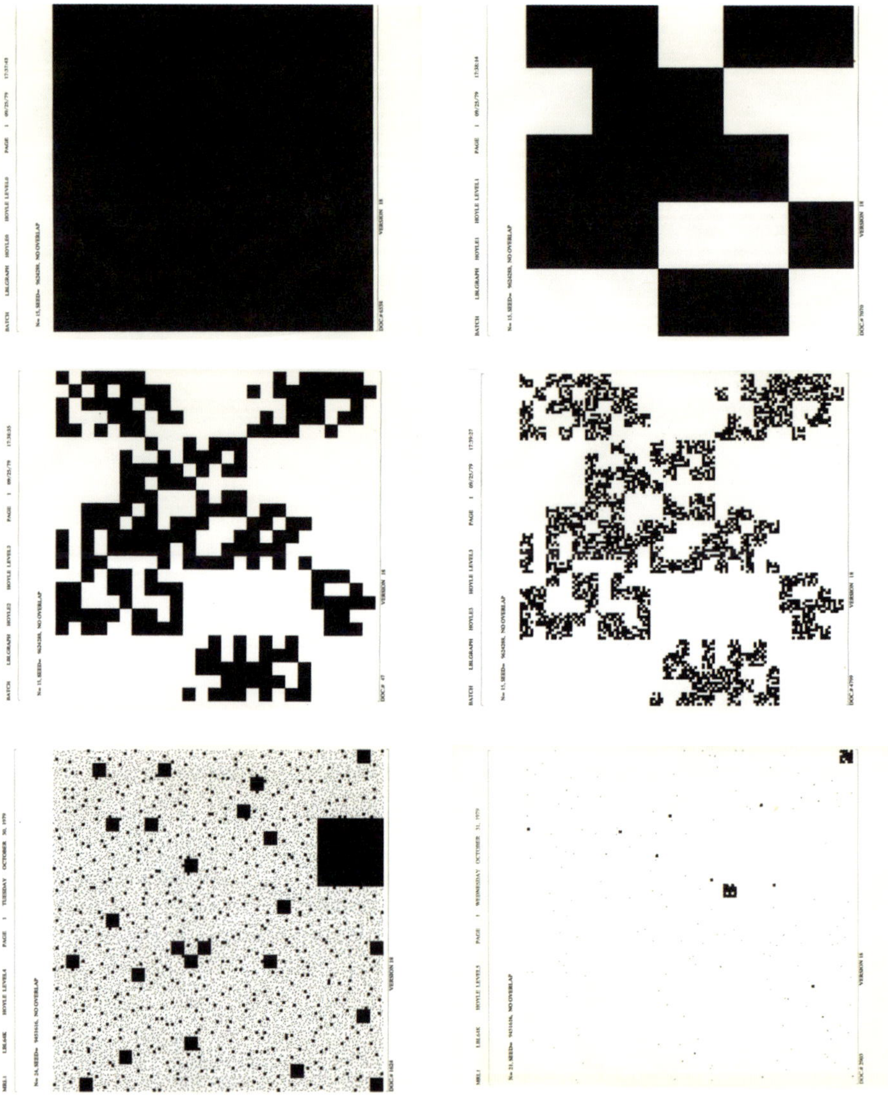

Cats. 26–31. Benoît Mandelbrot and Mark R. Laff, programmer. Implementation of Fred Hoyle's cosmological model, 1979. Computer-generated prints on photographic paper. Collection Aliette Mandelbrot.

Chains of Chance: A Convoluted History of Pattern Recognition

Wladimir Velminski

Picture Question

When black grids appear out of the void and lines suddenly change direction, chance is at work on a composition. Consider two pictures that could not be more different and yet at some level are also very similar: Both contain four squares with different patterns, but in the first one, highly varied angles fill the surfaces of the squares (fig. 1, left), while in the second, the surfaces are filled with rectangular pixels in varying shades of gray (see fig. 1, right). The first is an artwork by Frieder Nake, the second a "plate" by Benoît Mandelbrot illustrating a scientific model. What constitutes the essential identity of these pictures?

In the Face of the Irregular

It is well known that Benoît Mandelbrot begins the book *The Fractal Geometry of Nature* with a declaration. "There is a fractal face to the geometry of nature."[1] In this "scientific essay," he discloses that many forms of nature obey one and the same principle, namely self-similarity or, to use Mandelbrot's term, *scale invariance*. Based on this principle—according to which the infinite enlargement of a patterned object continuously reveals the original pattern repeated at ever greater scales—fractal geometry reveals its "hidden face."[2] The following is an attempt to retrace the path to fractals, which itself resembles the

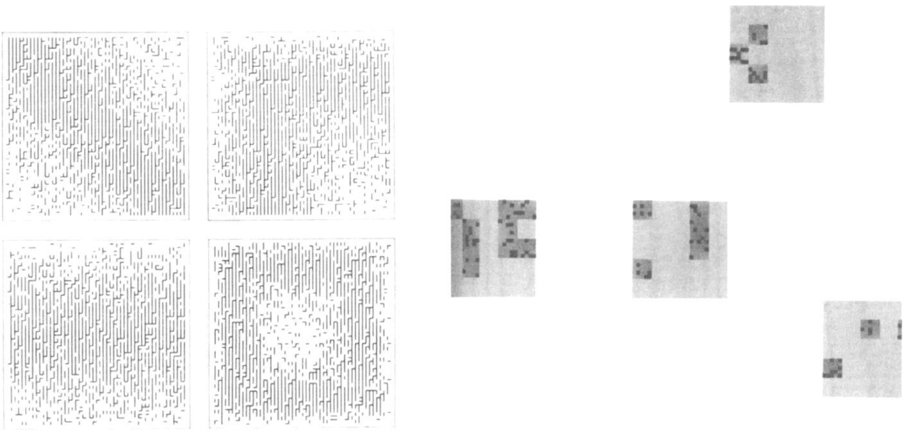

Fig. 1. (left) Frieder Nake, "Four Realizations of Walk-Through Rasters, from Series 2.1," 1966. From Frieder Nake, *Ästhetik als Informationsverarbeitung. Grundlagen und Anwendungen der Informatik im Bereich ästhetischer Produktion und Kritik* (Vienna: Springer, 1974), 236, illus. 5.5-5. With kind permission of Springer Science and Business Media. (right) Benoît Mandelbrot, "Implementation of Hoyle's Model (Dimension D=1) Using Random Curdling in a Grid," computer drawing. From Benoît Mandelbrot, *The Fractal Geometry of Nature* (New York: Freeman, 1982), 219.

curves analyzed in Mandelbrot's book. He cryptically alludes to the path in the book's epilogue, where he speaks of a land (which he does not name) with which he has had a lifetime involvement; of a review that he retrieved from the wastebasket of a "pure" mathematician; of regularities in word frequencies and the birth of mathematical linguistics; of accidents, probabilities, and chains—and of the fact that though all of these things belong to the past, he has not stopped sensing their aftereffects.[3] The path consisted of the use of certain conjectural techniques to derive order from chance processes. It was out of this that the "geometric face" of the theory of scaling emerged, that it "grew in importance and gave size to fractal geometry."[4]

The Instrument of Chance

On January 23, 1913, the Russian mathematician Andrei Andreyevich Markov (1856–1922) presented a meeting of the Physical-Mathematical Division of St. Petersburg's Academy of Sciences with the results of an experiment he had conducted on the letters of the Russian alphabet.[5] His goal was to establish the concept of a chain of chance variables, and the resulting article, "An Example of Statistical Investigation of the Text *Eugene Onegin* Concerning the Connection of Samples in Chains," was published a short time later in Volume 7 of the *Isvestia Imperatorskoj Akademii Nauk*. The radicality of Markov's experiment is fully apparent from his report.[6] The object of his statistical investigation had been *Eugene Onegin*, the Romantic novel in verse by the Russian literary giant Alexander Sergeyevich Pushkin (1799–1837) (fig. 2). Fully aware that it was the 200th anniversary of the publication of Jacob Bernoulli's classic treatise on probability theory,[7] Markov set out to analyze the Russian classic as a system of *linked quantities*. His gridlike blocks of sharply printed letters became the object of his theoretical investigation. Disregarding all punctuation marks and spaces between words, he arranged the first 20,000 letters from *Eugene Onegin* into 200 blocks of

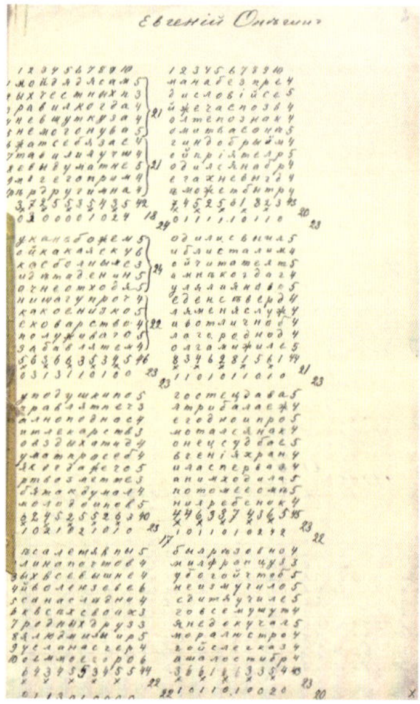

Fig. 2. Andrei A. Markov. "Pushkin in Block Form." From *Calculations on Eugene Onegin*, 1913. © St. Petersburg Branch of the Archive of the Russian Academy of Sciences, 2012. Holding 173, Register 1, File 118, Sheet 2.

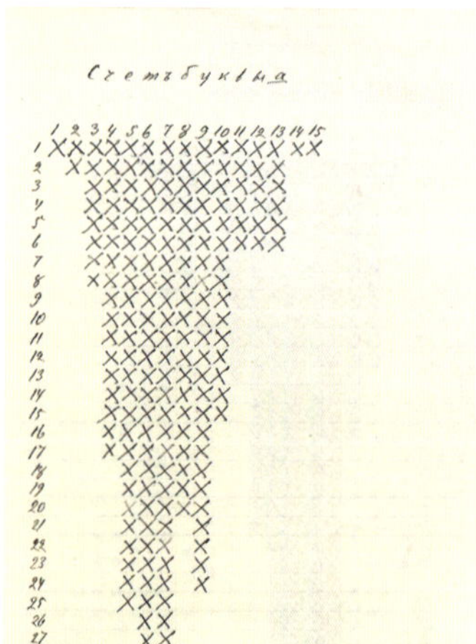

Fig. 3. Andrei A. Markov. "The Altered Count of the Letter A". From *Calculations on Eugene Onegin*, 1913. © St. Petersburg Branch of the Archive of the Russian Academy of Sciences, 2012. Holding 173, Register 1, File 118, Sheet 12r.

letters, each with ten columns and ten rows. From the interpretive section of Markov's report, it is clear that he wanted to produce a model from which one might derive the probabilities of the alternation between vowels and consonants that is typical of Russian as well as English and many other languages (fig. 3). He felt that it should be possible to predict the appearance of a given letter from the letters that had preceded it. In this way, language was transformed into an arrangement of links on paper that served as the basis of a new probability theory.

The Biometry of Literary Style

Markov's study did not wait long for a response. In 1915, in Volume 20 of the St. Petersburg *Isvestia Akademii Nauk, otdelenie russkogo jazyka* (section of Russian language), Nikolai Aleksandrovich Morozov (1854–1946) published an analysis of literary style. Morozov's "linguistic spectra"—the phrase appears in the title of his article—were intended to serve as "a way to distinguish plagiarisms from the genuine works of a given author," and their sophisticated linkings were to provide objective proof of the linguistic pattern system.[8] Morozov, a poet, writer, and linguist who would become an honorary member of St. Petersburg's Academy of Sciences, had acquired his scientific knowledge through lay experiments in cosmology, physics, chemistry, biology, geophysics, and meteorology, which had taught him that true accuracy in science is dependent entirely on statistics.[9] Morozov now wanted to use statistics as the basis for a precise, scientific analysis of language. He wrote: "If in nature and everyday life all manner of events appear at first to be coincidental, but after a certain number of repetitions take on a certain predictability, why cannot the same be true of language? The repetitions of words in our oral and written language must also demonstrate various probabilities."[10] The regularities of nature that Morozov had observed in heavenly bodies, mechanical processes, and the composition of matter he also saw in the way every author uses particular turns of phrase:[11] "Some writers frequently use the word 'which,' others don't like it, and replace it with the participial form of the verb, which is added at the end. Some use the particle 'between,' others write in place of it: 'amid' or 'within.'"[12]

Morozov postulated that all such distinctions in the structure of language can be traced back to "purely *mechanical causes*" consisting of "linked chains with internal and external linguistic influences" that "have long since become established in the realm of the unconscious."[13] For precisely this reason, Morozov accorded to particles a "switching function." According to his interpretation, such mechanical processes serve to regulate and organize our speech and the writing of poets.[14]

In order to analyze authorship questions, Morozov needed a method built on "reliable material." He therefore concentrated on linguistic chains, not on conceptual ones, which would "differ with every original writer."[15] Linguistic sequences are made up of individual elements "that in specific speech chains are not only typical of nearly all languages, but also appear in the same way in nearly all literary genres" and therefore can be termed "*functional* or *organizing* particles of human speech."[16] One such particle is the Russian negation "*ne*," which Morozov decided to study in the writing of nineteenth-century authors. To him, questions of authorship became a game of numbers:

> Count, and you will see that in every thousand words in Tolstoy this [negation] appears on average somewhat fewer than twenty times, in Pushkin and Gogol roughly twenty times, and in Turgenev a bit more than in his predecessors, in the realm of between twelve and thirty-five times per thousand words, always depending on the given writer's proclivity to negation.[17]

Morozov thus observed that the use of "the negative particle '*ne*' depends in large part on individual variation," and accordingly can be said to characterize the structure of each author's language.[18] This appears to contradict the proposition that language patterns can be predicted via the logic of the chain: In Morozov's system, the author's uniqueness is analogous to the uniqueness of the given linguistic work. His implicit criticism of Markov becomes all the more apparent when at the end of his introduction he states: "Unfortunately, you will not be able to arrive at any conclusion if all the words are written together without spaces and without punctuation."[19] Despite this

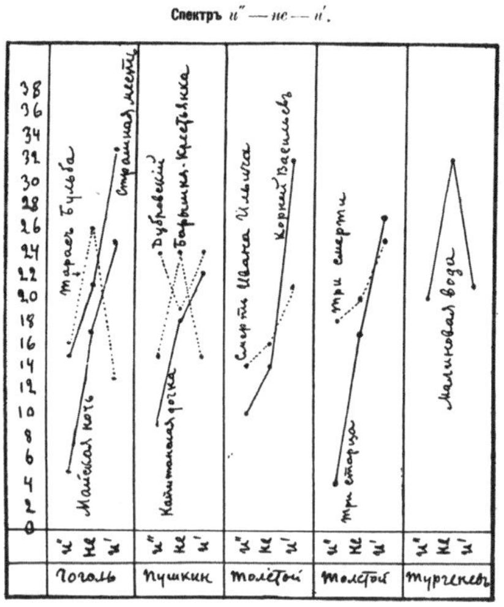

Fig. 4. "Linguistic Spectra." From N. A. Morozov, "Lingvističeskie spektry. Sredstvo dlja otličenija plagiatov ot istinnych proizvedenij togo ili drugogo izvestnogo avtora," *Isv. Akad. Nauk, otd. Russ. Jaz., Sankt Peterburg*, vol. 20 (1915), table IV, 108. © Moscow Branch of the Archive of the Russian Academy of Sciences, 2012.

objection, Morozov was directly indebted to Markov. He not only took up his predecessor's experiment, he, like Markov, was driven by a desire to visualize chance. Indeed, the graphics he produced in his pattern experiments led him to identify authors' textual quirks and to locate them within linguistic spectra (fig. 4). But though Morozov's research was based on pictures of chance, it was published as "linguistic analysis."[20] In order to differentiate his spectra, Morozov broke them down into prepositional, subjunctive, pronominal, and other syntactic groupings and superimposed the corresponding graphs. At this point, his picture took on an authority that seemed to exceed that of the linguistic expert. An overlay of the graphs made it clear that "with only a simple glance at the prepositional spectrum of any given work of Pushkin that was possibly erroneously attributed to Gogol we would have recognized the inaccuracy of that assertion and accordingly expressed the suspicion that the work was likely one of Pushkin's."[21] However, as a statistician, Morozov knew that only averages can be compared in this manner. His overlay graphs could not serve as incontrovertible proof. Even so, he initially concentrated on 1,000 words, from which he calculated a mean for the particles. He identified this mean as a natural value that determined what he called the author's "syllable physiognomy."[22] The spectra were sufficiently precise that they could be said to reveal the author's specific signature and set him apart from other writers. Here "physiognomy" is not merely a figure of speech: The visible countenance of language, represented by an overlay of graphs, became a mark of individuality, one with which it was possible to distinguish genuine authors from false ones. Chance is not without order. When Mandelbrot later felt that he was looking at the "fractal face of nature," he was translating this same experience into his theory and his pictures.

True to the principle of the chain reaction, Markov, the man who initiated the notion, responded to Morozov's provocation a year later. He asserted that Morozov's idea was not fundamentally wrong, but that its implementation should by no means be based on a subjective sense of literary style, and that counting a mere 1,000 words was not enough. He thereby dismissed Morozov's theory of the spectra embodied in the example of the negation as unsubstantiated by either literature or mathematics. Markov, who never tired of counting, used Morozov's own method against him, raising threefold the count of negations in the given authors and thereby demonstrating the inadequacy of Morozov's linguistic spectra as opposed to his own chain patterns. Markov insisted that precise, scientific analysis of language, as undertaken by Morozov, would have to rely on a mathematical-statistical foundation, one that presupposed the counting of several thousand sentences, words, syllables, and letters.[23] Since the invention of the computer, automatic speech recognition has been achieved on just such a basis, its fundamental problem being the establishment of a connection between the sequence of acoustical sounds and the words that correspond to it.

The Principle of Least Effort

The pattern systems of the 1910s, which were based on the frequencies of letters and individual words, were again taken up in the 1930s by George Kingsley Zipf. A lecturer at Harvard, Zipf expanded Markov and Morozov's methods in his own research by investigating the frequency of words in various languages, information that he used to recognize new speech patterns.[24] Zipf arranged the words in a language according to their frequency of appearance, and he managed to derive from this various mathematical constants. For example, the more frequently a word is used, the shorter it is. A word's age also has to do with how often it is used.[25] These laws would eventually find their way into the major study on communication that he published in 1949 under the title *Human Behavior and the Principle of Least Effort*.[26] There Zipf applied his mathematical findings to people's speech habits: Frequently-used words are shortened in their sound structure whereas greater length is tolerated in less common words. It was from this observation that Zipf derived his model of communication: Every speaker wishes to communicate his message with the least possible effort. Yet he encounters a difficulty, for the listener or reader on the other end of the channel wants to decipher that message with the least possible effort.[27] It is only when these two opposing interests are brought into balance that communication is possible.

Among other things, it was a review of this book that led Mandelbrot to fractals: "This empirical regularity concerning word frequencies had come to my attention through a book review. The event seems too symbolic to be true, but the review in question had been retrieved from a 'pure' mathematician's wastebasket, for light reading on the Paris subway."[28] The pure mathematician was Mandelbrot's uncle, Szolem Mandelbrojt, with whom Mandelbrot frequently stayed between 1945 and 1947 while studying at the École polytechnique and whom he visited regularly while working on his doctorate. The reviewer was in fact the mathematician Joseph Leonard Walsh, who in 1949 discussed the study of his Harvard colleague Zipf in *Scientific American*. As chance would have it, in his dissertation, "Contribution à la théorie mathématique des jeux de communication,"[29] Mandelbrot was combining previously known theories of linguistic analysis with Norbert Wiener's cybernetics and John von Neumann's game theory, which were also in vogue at that time. By analyzing language from a macrolinguistic perspective rather than a microlinguistic one, Mandelbrot hoped his study would make him "the Kepler of mathematical linguistics."[30] He placed himself in the Markov tradition but pointed out that the Russian's theory had meanwhile been expanded beyond dependent chance quantities by a series of new concepts and interpretations that had changed the shape of his chains. This further development was mainly owing to Claude E. Shannon, who in 1947 showed that the Markov chain was still useful as a mathematical model for the theory of information transmission. As Shannon's illustration shows, code elements can quantify the information to be communicated with numbers (fig. 5, top). Shannon relied on the theory of Markov chains. "The main point at issue," he wrote, "is the effect of statistical knowledge about the source in reducing the required capacity of the channel, by the use of proper encoding of the information"[31] (see fig. 5, bottom). Mandelbrot, in turn, expanded this model for his own purposes; what was crucial in its arrangement of signs was the introduction of a space sign, one that cannot appear within a word. It was an element that mathemati-

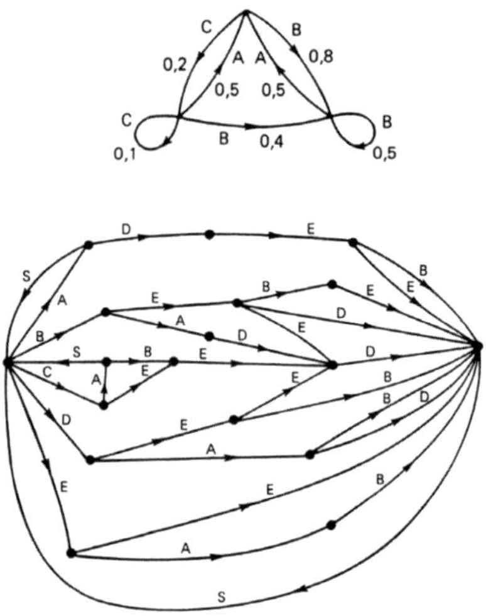

Fig. 5. "Markov Processes." From Claude E. Shannon, *Mathematical Foundations of Information Theory* (New York: Dover, 1976), 16, illus. 4–5.

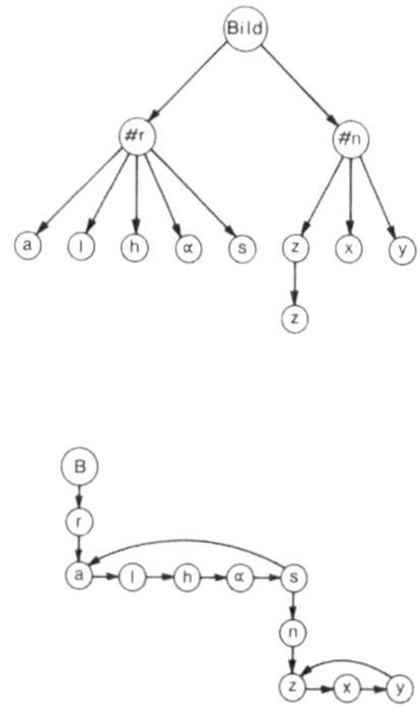

Fig. 6. "Markov Processes." From Frieder Nake, *Ästhetik als Informationsverarbeitung. Grundlagen und Anwendungen der Informatik im Bereich ästhetischer Produktion und Kritik* (Vienna: Springer, 1974), 235, illus. 5.5-5. With kind permission of Springer Science and Buisness Media.

cians like Markov and Shannon had omitted in their language patterns but that linguists like Morozov and Zipf had reintroduced. Merely marking the boundaries between words, the sign is not dependent on adjacent letters.[32] As a result, the individual word becomes the central interval of chance,[33] and from the point of view of probability theory, it is revealed that language shows a high degree of self-similarity. Mandelbrot recognized in this the scale invariance that would become one of the fundamental principles of his later work.

One also sees this in his implementation of the cosmological model of Fred Hoyle, which was based on a chance curdling of forms in a grid (see fig. 1, right). Regardless of where and in which shade of gray the grids appear, they are determined by the probabilities of transition. The same is true of the angles in Frieder Nake's *Walk-Through Raster* series (see fig. 1, left): Here it is a Markov chain of chance events that determines where a given transition will occur (fig. 6).[34] In the case of Mandelbrot's Hoyle visualizations, the chance events are determined by means of a transition probability, be it the flip of a coin, a given birthrate, or the calculated distribution of galaxies that influence degrees of density. The codes and the sources are thus linked together. It was for this reason that Mandelbrot wrote at the end of *The Fractal Geometry of Nature*: "At a critical point in my life I read a wise review of *Human Behavior* by the mathematician J. L. Walsh. By only mentioning what was good, this review influenced greatly my early scientific work, and its indirect influence continues. Therefore, I owe a great debt to Zipf through Walsh."[35] Without the attempts to make visible the linkings of chance that began with Markov, Mandelbrot would not have been able to recognize the "fractal face of nature."

1 Benoît Mandelbrot, *The Fractal Geometry of Nature* (New York: Freeman, 1982), 3.
2 Ibid., 4.
3 Ibid., 422–24.
4 Ibid., 424.
5 For a detailed description of the experiment and its ramifications, see Philipp von Hilgers and Wladimir Velminski, eds., *Andrej A. Markov. Berechenbare Künste. Mathematik, Poesie, Moderne* (Zurich: diaphanes, 2007).
6 Aleksandr Chuprov points to the importance of Markov to his theory. In a posthumous review that discusses Markov's fourth edition of his probability calculation (Markov 1924), Chuprov suggests to his longtime correspondent that his general approach relied on complex avenues, and that Markov "assumes great importance when he is taken up by a broad scientific background, which he deserves." See Hilgers and Velminski, eds., *Andrej A. Markov*, 32–39.
7 Markov was very much involved in the organization of its celebration. See the letter from Markov to Chuprov from January 15, 1913, in A. A. Markov, "Primer statističeskogo issledovanija nad tekstom 'Evgenija Onegina,' illyustrirujuščij svjaz' ispytanij v zep," *Izv. Imp. Akad. Nauk, Sankt Peterburg*, ser. 8, vol. 7, no. 3 (1913), 153–62.
8 N. A. Morozov, "Lingvističeskie spektry. Sredstvo dlja otličenija plagiatov ot istinnych proizvedenij togo ili drugogo izvestnogo avtora," *Isv. Akad. Nauk, otd. Russ. Jaz.*, Sankt Peterburg, vol. 20 (1915), 93–134. See also Wladimir Velminski, "Der Speck am Text. Die Entstehung der Genetik aus der Berechenbarkeit der Literatur," in Hilgers and Velminski, eds., *Andrej A. Markov*.
9 N. A. Morozov, *Povest'moej žizni* (Moscow: Izd-vo Akademii nauk SSSR, 1961), 379.
10 Morozov, "Lingvističeskie spektry," 100.
11 See N. A. Morozov, *Die Evolution der Materie auf den Himmelskörpern: Eine theoretische Ableitung des periodischen Systems* (Dresden: T. Steinkopff, 1910) and Morozov, *Die Offenbarung Johannis: eine astronomsich-historische Untersuchung* (Stuttgart, 1912).
12 Morozov, "Lingvističeskie spektry," 94.
13 Ibid.
14 Ibid.
15 Ibid., 101.
16 Ibid.
17 Ibid.
18 Ibid., 102.
19 Ibid., 100.
20 Ibid., 103
21 Ibid., 105.
22 Ibid., 109.
23 A. A. Markov, "Ob odnom primenenie statističeskogo metoda," *Izv. Imp. Akad. Nauk, Sankt Peterburg*, ser. 6, Vol. 10, no. 4 (1916), 239–42.
24 See George Kingsley Zipf, *Selected Studies of the Principle of Relative Frequency in Language* (Cambridge, MA: Harvard University Press, 1932); and Zipf, *The Psycho-Biology of Language: An Introduction to Dynamic Philology* (Cambridge, MA: Harvard University Press, 1935).
25 See also David Crystal, ed., *The Cambridge Encyclopedia of Language* (Cambridge: Cambridge University Press, 1987); and C. Prün, "G. K. Zipf's Conception of Language as an Early Prototype of Synergetic Linguistics," in *Journal of Quantitative Linguistics* 6 (1999).
26 George Kingsley Zipf, *Human Behavior and the Principle of Least Effort: An Introduction to Human Ecology* (Cambridge, MA: Harvard University Press, 1949), 24–35.
27 See Zipf, *Human Behavior and the Principle*, 33–35.
28 Mandelbrot, *The Fractal Geometry of Nature*, 422.
29 Benoît Mandelbrot, "Contribution à la théorie mathématique des jeux de communication" (PhD diss., l'Université de Paris, 1953), 2, 10.
30 A. Barcellos, "Interview of B. B. Mandelbrot," in *Mathematical People* (Boston: Birkhäuser, 1984), 3.
31 Claude E. Shannon and Warren Weaver, *The Mathematical Theory of Communication* (Urbana: University of Illinois Press, 2009), 10.
32 See M. Bullynck, "Markovketten—Variationen im Computer," in Hilgers and Velminski, eds., *Andrej A. Markov*, 162–64.
33 Bullynck, "Markovketten," 162–64.
34 Frieder Nake, *Ästhetik als Informationsverarbeitung. Grundlagen und Anwendungen der Informatik im Bereich ästhetischer Produktion und Kritik* (Vienna: Springer, 1974), 235.
35 Mandelbrot, *The Fractal Geometry of Nature*, 404.

Found
Images V

Scaling as Pictorial Strategy: Scale Ambivalence and Openness to Interpretation

Nina Samuel

"Roughly stated, fractal geometry is the study of scale-invariant roughness."[1] Just how Mandelbrot preferred to translate this central quality of fractals into images is evident from several finds from his office: A black blot and three amorphous, irregular agglomerations whose material makeup is difficult to identify are pasted together on a sheet of blue-lined paper (cat. 1). The picture shows the final layout for the two-page plate "Fractal Flake and Its Shadow" as published in Mandelbrot's 1977 book *Fractals*.[2] The term "flake" refers to experiments by the French physicist Jean Perrin, who studied Brownian motion—the irregular and temperature-dependent molecular movement of the tiniest particles of ash and dust—at the beginning of the twentieth century.[3] In 1975, Mandelbrot declared Brownian motion to be the "prototype of a fractal,"[4] and it proved to be one of his main research subjects. He produced a vast number of graphics in order to study Brownian motion's characteristics and the possible ways in which it could be illustrated. In a note from July 1976, for example, his colleage Richard Voss informed him that a new batch of "fractal flakes" had been completed "to (hope-fully) meet your minimum daily requirements" (cat. 2). The note includes technical details relating to the manner in which they were produced as well as eight slightly different "flakes" that show that minimal variations in shape or shading and lighting were already crucial (cats. 3–10).

The impression that these mathematical objects have been "illuminated" is particularly enhanced where some of the flakes are accompanied by huge shadows. The image was treated as if it were in fact a real object that one could approach or back away from; turn; or, indeed, illuminate. This was consistent with Mandelbrot's general concept of mathematical objects: "I have the feeling that in a way I accomplished so many different things only by simply looking at an object which I was not familiar with and manipulating it as an object."[5]

The way such "objects" are presented on the layout paper for *Fractals* (see cat. 1) is particularly significant. Insofar as all three of them are subjected to the same lighting conditions, one might see them as occupying a common picture plane. However, the explanatory text makes it clear that they are meant to represent three different enlargements. Whereas the flake in the lower right is an overall view from a distance, the upper images are supposed to represent close-ups. At the same time, the range of their possible interpretations seems virtually endless. Mandelbrot saw not only similarities to a "small asteroid," but also to mysterious creatures: "An earlier version of a portion of this figure was reminiscent of one of the better known 'photographs' of the Loch Ness monster."[6] Whether invoking astronomy or the realm of fable, Mandelbrot deliberately played up such leaps between levels of visibility, orders of magnitude, and possible identifications as being a central paradigm in the design of fractals. From the object itself, seemingly floating freely in space, one could determine neither its scale nor its essential nature, whether it was organic or inorganic. And the picture was not meant to provide such information but only to stir the imagination and entertain diverse interpretations. Mandelbrot's "flakes" are perfect examples of visualizations of the fractal characteristic known as "scaling," as they are deliberately ambiguous and devoid of any indication of size.

Most of the phenomena Mandelbrot studied follow this scale-invariant structure, and did so long before he began using computer graphics.[7] In time, the aesthetic principle of scale-invariance became such an effective organizational tool that it not only structured his research but also his thinking outside the field of mathematics: He organized his life according to the concept of fractal self-similarities.[8] It also became a kind of "stylistic" distinction, the mediator between nature, art, and science.[9] The parallels to art history he invoked were only one aspect of his attempt to apply his fractal geometry across separate disciplines as proof of the universality of his aesthetics: "I wanted to make the point that fractals are one of the things that humanity has always known. But who was humanity which we have evidence of? They are mostly painters and some sculptors!"[10]

1 Benoît Mandelbrot, *Gaussian Self-Affinity and Fractals* (New York: Springer, 2002), 9. Self-similarity is the geometrically stricter variant of the overriding principle of scale-invariance.
2 Benoît Mandelbrot, *Fractals: Form, Chance, and Dimension* (San Francisco: Freeman, 1977), 8.
3 See Jean Perrin, "Mouvement Brownien et Molécules," *Journal de Physique Théorique et Appliquée* 9 (1910). For Brownian motion, see Charlotte Bigg, "A Visual History of Jean Perrin's Brownian Motion Curves," in *Histories of Scientific Observation*, ed. Lorraine Daston and Elizabeth Lunbeck (Chicago: University of Chicago Press, 2011). For the experimental study of Brownian motion as part of the pre-history of fractals, see Jean-Luc Chabert, "Un demi-siècle de fractales: 1870–1920," *Historia Mathematica* 17 (1990): 345–47.
4 Benoît Mandelbrot, *Les objets fractals: forme, hasard et dimension* (Paris: Flammarion, 1975), 10.
5 Benoît Mandelbrot, interview by author, Cambridge, MA, April, 2008.
6 Mandelbrot, *Fractals*, 8.
7 Though the term had yet to be invented, Mandelbrot first found "self-similar" order in technical interferences, which he and a group of engineers at IBM were tasked with eliminating in electrical phone lines in the late 1950s. By 1965, at the latest, he had begun employing the term to describe the behavior of dynamic systems as they are describable by equations. See Mandelbrot, "Self-Similar Error Clusters in Communication Systems and the Concept of Conditional Stationarity," *IEEE Transaction on Communications Technology* 13 (1965). Mandelbrot claims to have coined the term at Harvard in 1964; see Mandelbrot, "Comment j'ai découvert les fractales," *La Recherche* 17 (March 1986): 422.
8 "[M]y life had followed a fractal orbit," Mandelbrot, quoted from A. Barcellos, "Interview of B. B. Mandelbrot," in *Mathematical People* (Boston: Birkhäuser, 1984), 208.
9 Mandelbrot, "Fractals and an Art for the Sake of Science," *Computer Art in Context: SIGGRAPH '89 Art Show Catalogue*, supplementary issue of *Leonardo* (1989): 21–24. See also, Mandelbrot, "Scalebound or Scaling Shapes: A Useful Distinction in the Visual Arts and in the Natural Sciences," *Leonardo* 14 (1981).
10 Benoît Mandelbrot, interview by author, Cambridge, MA, April, 2008. The best-known work of non-Western art to which he referred is surely *The Great Wave Off Kanagawa* (1829–32) by Katsuchika Hokusai; see Benoît Mandelbrot, *The Fractal Geometry of Nature* (New York: Freeman, 1982), color plate C16.

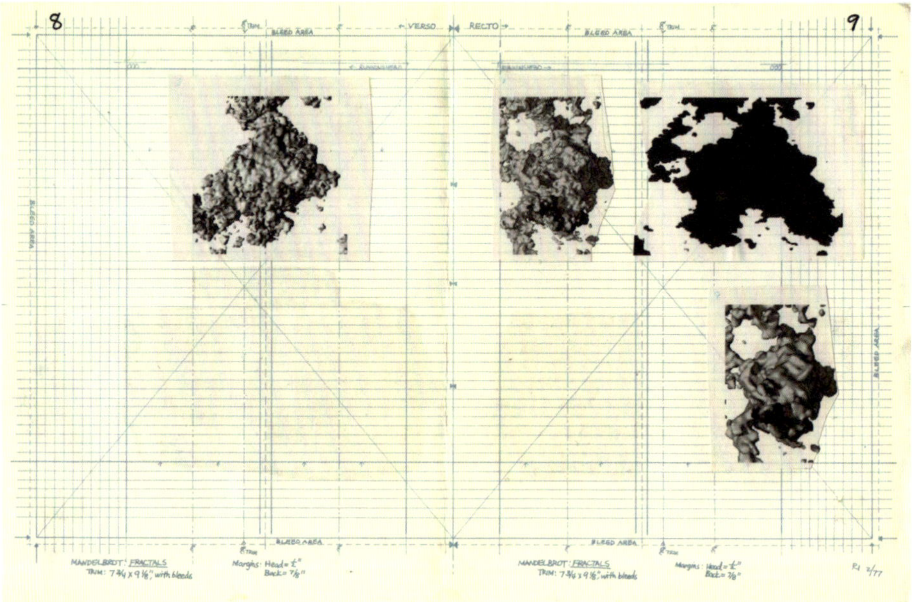

Cat. 1. Benoît Mandelbrot and Richard Voss. "Fractal Flake and Its Shadow," ca. 1976–77. Mockup for double page printed in *Fractals*, 1977. Collage on paper. Collection Aliette Mandelbrot.

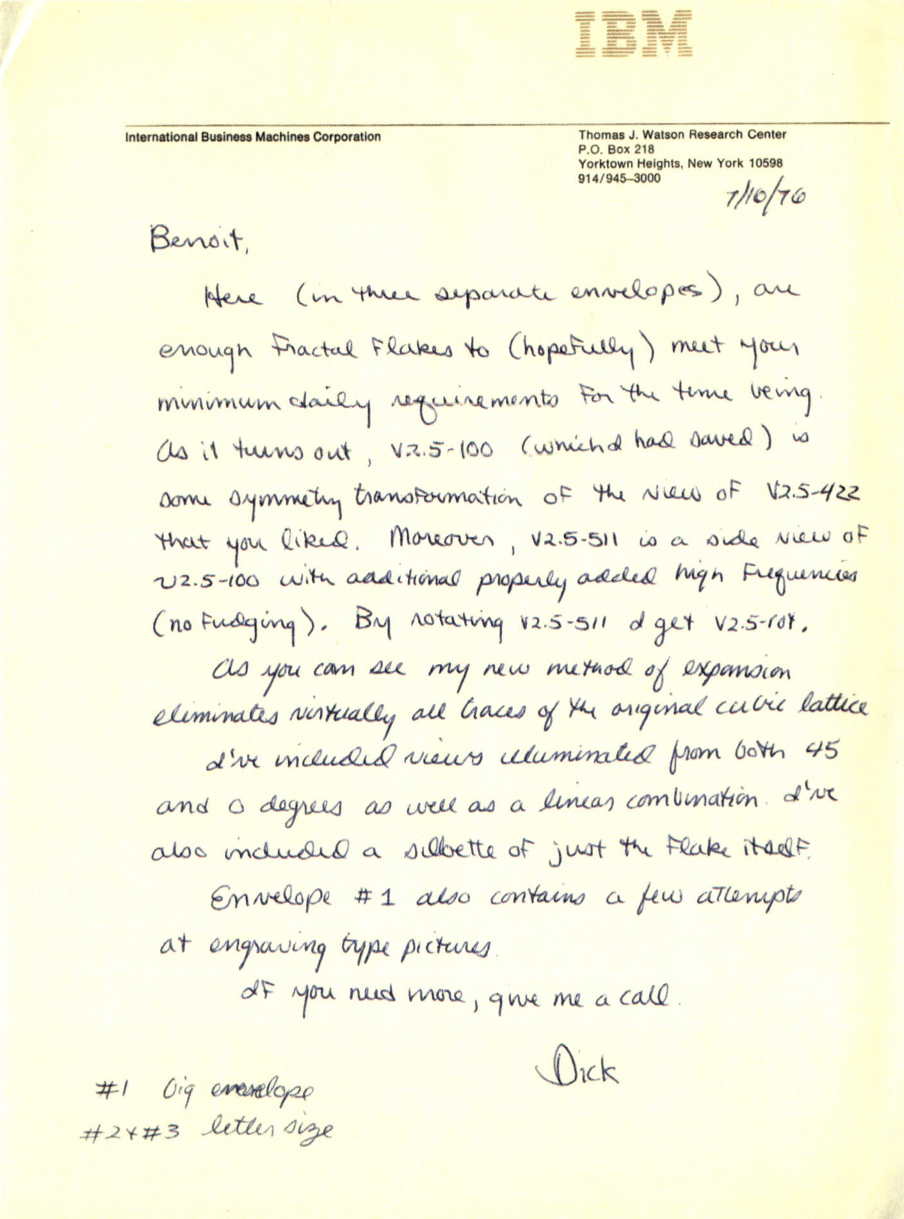

IBM

International Business Machines Corporation

Thomas J. Watson Research Center
P.O. Box 218
Yorktown Heights, New York 10598
914/945-3000

7/10/76

Benoit,

Here (in three separate envelopes), are enough Fractal Flakes to (hopefully) meet your minimum daily requirements for the time being. As it turns out, V2.5-100 (which I had saved) is some symmetry transformation of the view of V2.5-422 that you liked. Moreover, V2.5-511 is a side view of V2.5-100 with additional properly added high frequencies (no fudging). By rotating V2.5-511 I get V2.5-104.

As you can see my new method of expansion eliminates virtually all traces of the original cubic lattice

I've included views illuminated from both 45 and 0 degrees as well as a linear combination. I've also included a silhouette of just the flake itself.

Envelope #1 also contains a few attempts at engraving type pictures.

If you need more, give me a call.

Dick

#1 big envelope
#2 & #3 letter size

Cat. 2. Richard Voss. Letter addressed to Benoît Mandelbrot, July 10, 1976. IBM. Collection Aliette Mandelbrot.

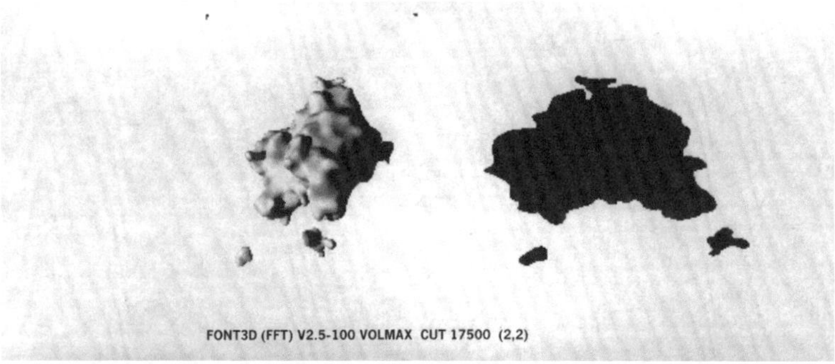

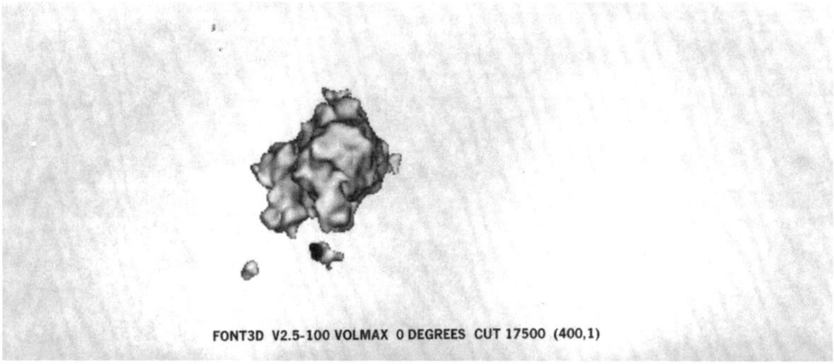

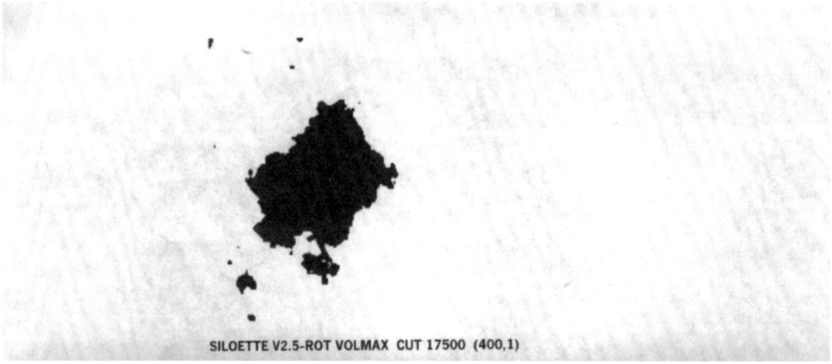

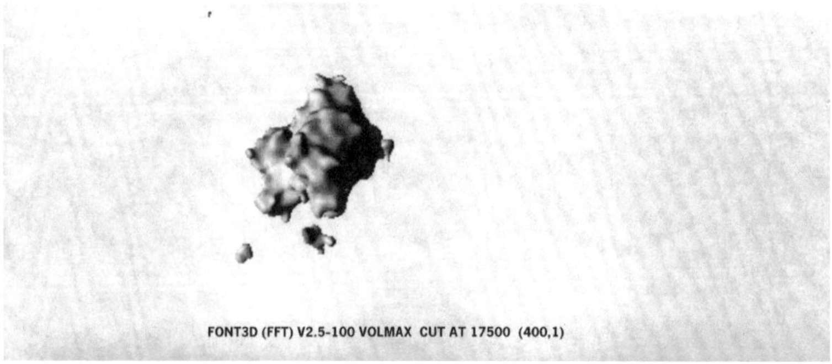

Cats. 3–10. Benoît Mandelbrot and Richard Voss. "Fractal Flakes," series of eight, ca. late 1970s. Computer-generated prints. Collection Aliette Mandelbrot.

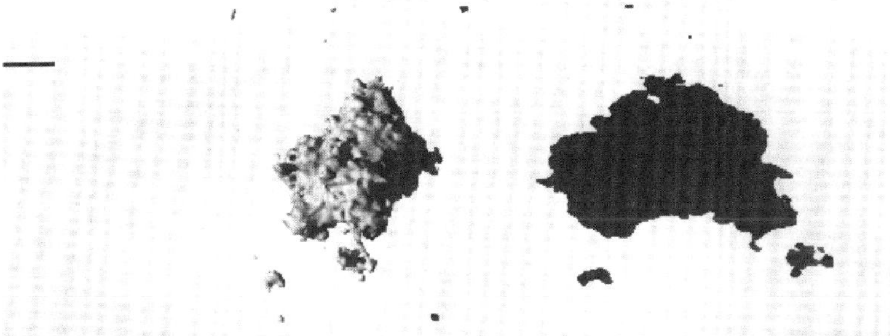

FONT3D V2.5-ROT VOLMAX 45 DEGREES CUT 17500 (2,2)

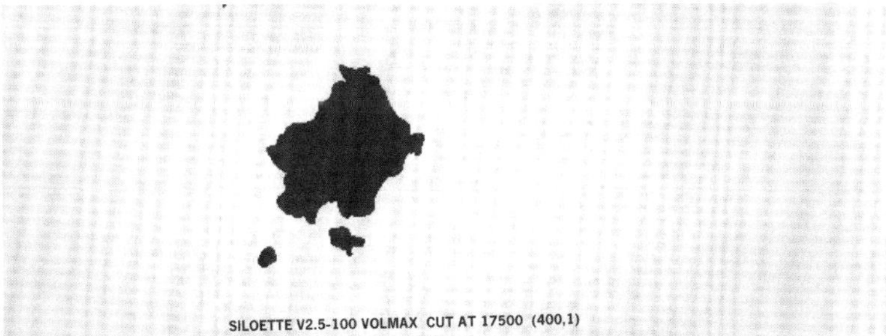

SILOETTE V2.5-100 VOLMAX CUT AT 17500 (400,1)

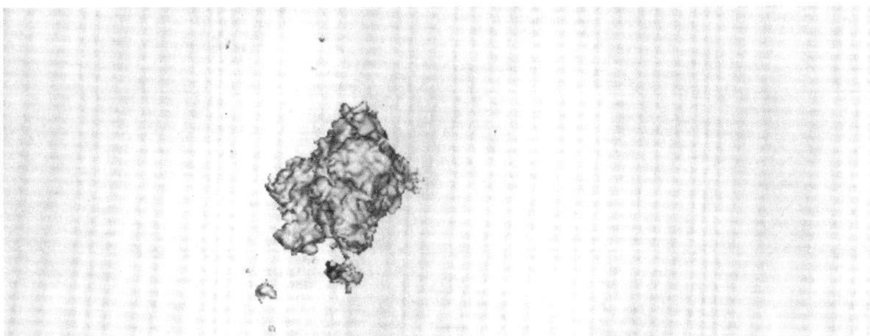

FONT3D V2.5-ROT VOLMAX 0 DEGREES CUT 17500 (400,1)

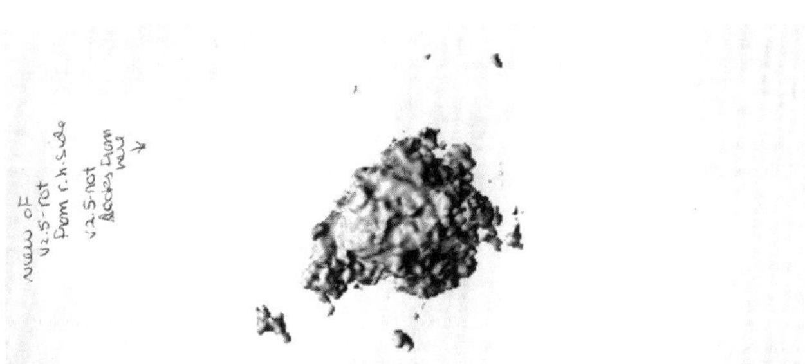

FONT3D-0 V2.5-511 VOLMAX 0 DEGREES CUT 17500 (400,1) RF 0

Scalebound Bauhaus

Juliet Koss

In 1981, Benoît Mandelbrot published a brief essay in the journal *Leonardo* entitled "Scalebound or Scaling Shapes: A Useful Distinction in the Visual Arts and in the Natural Sciences." Its argument, presented with apparent objectivity and helpfully summarized within its own title and subtitle, is that the concepts of "scalebound" and "scaling" that he had developed for his work on "shapes that I call fractal sets" might also prove productive in other fields, not only within the sciences but also (and more provocatively) in the arts.[1] In setting out his argument, Mandelbrot used the term "*scalebound* to denote any object, whether in nature or one made by an engineer or an artist, for which characteristic elements of scale, such as length and width, are few in number and each with a clearly distinct size." In his view, such objects did not achieve aesthetic clarity but instead were merely simplistic; by way of illustration, he provided the example of "a Bauhaus style, glass cube–type building," an abstract architectural category the fundamental scale elements of which—height, breadth, depth—were so clearly legible as to render the overall design simply banal.

The particular Bauhaus culprit whom Mandelbrot mentioned was neither Walter Gropius, the founding director of the school in 1919, nor Hannes Meyer, its director from 1928–30, but rather Ludwig Mies van der Rohe, its third and final director before the school's closure in Berlin in 1933. Yet with no specific buildings or designs offered as evidence, the phrase "glass cube–type building" remained vague, potentially encompassing anything from Mies's Farnsworth House (a one-room weekend home built outside Chicago in 1951) to the Seagram Building, the skyscraper completed seven years later on Park Avenue in New York City (fig. 1). Mandelbrot affixed

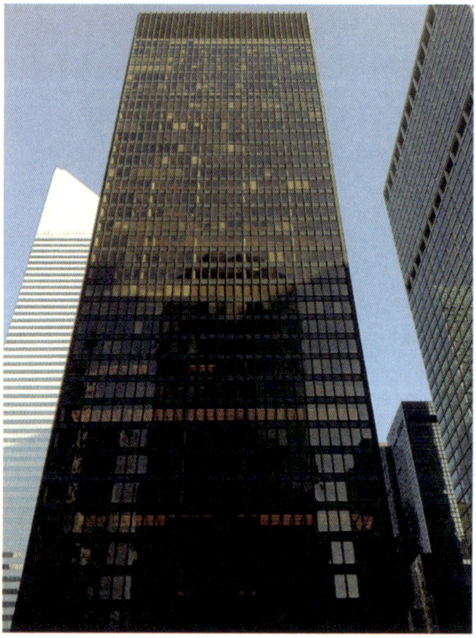

Fig. 1. Ludwig Mies van der Rohe. Seagram Building, New York, 1958, photographed 2012. Courtesy Juliet Koss.

the label "scalebound" to the architect's entire oeuvre, yet the accuracy of this characterization was perhaps irrelevant, given that "Mies" had by 1981 become shorthand for the postwar steel-and-glass-box architecture he had inspired and his followers produced: geometric skyscrapers that had proliferated in urban areas across the United States and Europe. Such buildings were clearly legible not only as architectural forms but also, to many viewers, as emblems of corporate functionalism—repetitive, soulless postwar constructions. In the context of an argument about the legibility of scale, Mandelbrot's reluctance to identify his architectural referents indicates, perhaps above all else, the malleability of the notion of the scalebound.

Whereas the relationship between the elements of a scalebound object were easily apprehended, Mandelbrot asserted, an object involving scaling possessed a structure so complex as to be nearly illegible, visually and conceptually. Whether designed by humans or produced by nature, such an object, he argued, "includes as its defining characteristic the presence of very many different elements whose scales are of any imaginable size. There are so many different scales, and their harmonies are so interlaced and interact so confusingly, that they are not really distinct from each other, but merge into a continuum."[2] Significantly, visual confusion—far from being lamentable—offered a welcome complexity, providing evidence of creative achievement or the inherent sophistication of certain natural forms. Examples for the category of scaling were also vague, ranging from "Occidental traditional compositions of visual art, as exemplified by representational figurative paintings and television productions and traditional architecture of the beaux-arts kind" to "the shapes of coastlines and the courses of rivers."[3] Scaling could occur in all artistic disciplines, media, nations, and historical eras.

Mandelbrot valued perceptually confusing images, objects, and natural formations, and his embrace of complexity extended to his own rhetoric. Indeed, the critique of the scalebound is itself neither immediately discernible nor entirely straightforward in his essay, the opening sentences of which might even be taken to set the stage for a defense of architectural modernism:

> It is often said that 20th-century "modern" buildings are sterile, not built to human scale and, in fact, unnatural. The more I ponder such statements and their variants, the more elusive I find their meaning and the more I feel that a discussion of the logic and the aesthetics of the notion of scale need to be addressed.[4]

One might expect, after such a beginning, an avowal of the fundamental naturalism of modern architectural forms, or perhaps an attempt to situate machine-age designs in the context of the morphological abstractions of the natural world. The mathematician Mandelbrot, after all, has stated that the "meaning" of the common critique of modern architecture is "elusive" and had mentioned logic and aesthetics in one breath; one might now easily imagine him addressing "the notion of scale" in order to unite these disparate elements and to promulgate, in the face of widespread criticism, the aesthetic logic of modern architecture.

Natural forms and modern architecture are indeed soon linked in Mandelbrot's text, but not in order to endorse either scalebound objects or modern design. Rather, much like the essay title, the comparison functions primarily to demonstrate a veneer of objectivity regarding the two terms under discussion—"scalebound" and "scaling"—each of which is linked to specific architectural models and to natural forms that range from the infinitesimal to the infinitely large. "One of my conclusions," Mandelbrot indicated in his essay's first paragraph,

> is that it is fruitful to call Mies van der Rohe's buildings scalebound—a term a physicist would use to describe a flawless crystal and the solar system—and to call the Paris Opera House a scaling building—the term scaling also being applicable to typical views of the Alps and to the visual characteristics of many other objects in nature, some of them visible (large or small) and others invisible to the naked eye.[5]

And who might imagine, following this seemingly balanced explanation, that such comparisons could function to bolster a critique of the modernist box, to formulate a logical demonstration of its aesthetic inferiority, and, ultimately, to justify the author's own taste in architecture?

Central to Mandelbrot's distinction between scalebound and scaling objects was the role of human perception, particularly as affected by the viewer's own physical size or distance from the object. Whereas the scale elements of scalebound objects remain visually consistent and clearly legible from any distance, he maintained, those of scaling objects appear to shift according to the proximity of the object and the viewer's size—to such an extent that it becomes impossible, or at least unproductive, to discuss the object in terms of scale. "For practical purposes," Mandelbrot wrote, "a scaling object does *not* have a scale that characterizes it. Its scales vary also depending upon the viewing points of beholders. The same scaling object may be considered as being of a human's dimension or of a fly's dimension."[6] If the scale elements of a Miesian skyscraper were immediately clear from any distance—and therefore to be considered unimaginative and dull—those of Alpine mountain ranges as well as "many other objects in nature," large and small, were indeterminate and therefore inherently more interesting.

Remarkably, however, the scale elements of the building that Mandelbrot presented as the architectural emblem for the notion of scaling, the Paris Opera House, remain as clear from a distance as those of any modernist box (fig. 2). Designed by Charles Garnier and completed in 1874, the ornate Beaux-Arts building possesses a main façade with architectural hierarchies designed to be visible from a great distance at the end of the grand Avenue de l'Opéra. In approaching the building, the attentive viewer finds its architectural proportions reproduced at the level of the individual façade segments, each one characterized by an insistent symmetry and an emphasis on visual balance and proportion.[7] Mandelbrot's claims that the building "does not have a scale that characterizes it" and that its "scales vary also depending upon the viewing points of beholders" thus reveal as much about his own weakness at parsing architectural form as about the building under discussion.

If the phrase "a Bauhaus style, glass cube–type building" had been inspired by the designs of epigonic Miesians—if not those of Mies himself—it also evokes an earlier, iconic Bauhaus building: the Haus am Horn, an experimental single-family home designed by Georg Muche and Adolf Meyer and constructed (under the supervision of Gropius's office) for the Bauhaus exhibition held that year in Weimar. Its modularity visible in its architectural plan, the house may be understood as an arrangement of two architectural boxes: a taller one containing the central living space and a shorter, broader one surrounding it, partitioned into small rooms—kitchen, dining room, three bedrooms, and a bathroom—around

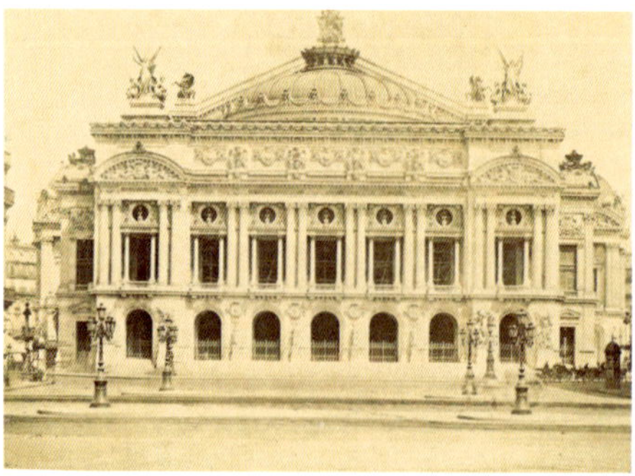

Fig. 2. Charles Garnier. Paris Opéra, front façade, 1876. Photograph. From *L'Opéra de Paris construit par M. Charles Garnier: Architecture, Sculpture, Décorative: Extérieur & Intérieur* (Paris: E. Bigot, [ca. 1876]). (92.R.7), The Getty Research Institute, Los Angeles.

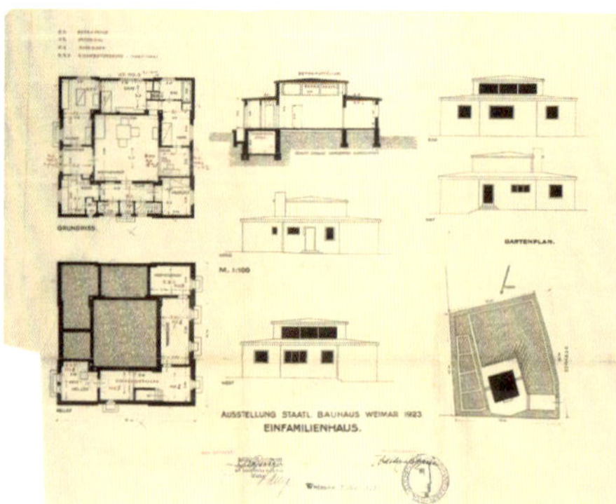

Fig. 3. Georg Muche and Walter Gropius. Architectural plan for the Haus am Horn, Weimar, 1923. Photocopy, ink, pencil. Klassik Stiftung Weimar Museums, L 2108a.

the perimeter (fig. 3). To use Mandelbrot's terms, the house would appear from the exterior to be scalebound; its "characteristic elements of scale, such as length and width, are few in number and each with a clearly distinct size."[8]

Easily legible on its exterior, the scale elements of the Haus am Horn are also echoed in miniature within its walls. Assigned the task of outfitting the children's room, the Bauhaus student Alma Buscher designed a cupboard that could be used for storing toys and books, presenting puppet shows, or (as the advertising copy from 1923 sweetly explains) "playing, sitting, and traveling" (fig. 4).[9] Such cubic furniture designs could be adapted for different uses as children grew older—for the more grown-up pleasures of reading, for example, or at least looking at a picture book of shapes, in an endless loop of modular modern design. Playing among its modular furniture in this carefully designed room, children could also enjoy the "Ship" Building Toy that Buscher created that year (fig. 5). Its modular blocks operating as miniature versions of the furniture designs in the modern house in which it could be configured and reconfigured, this toy was the smallest component within what Mandelbrot might have termed a scaling building.

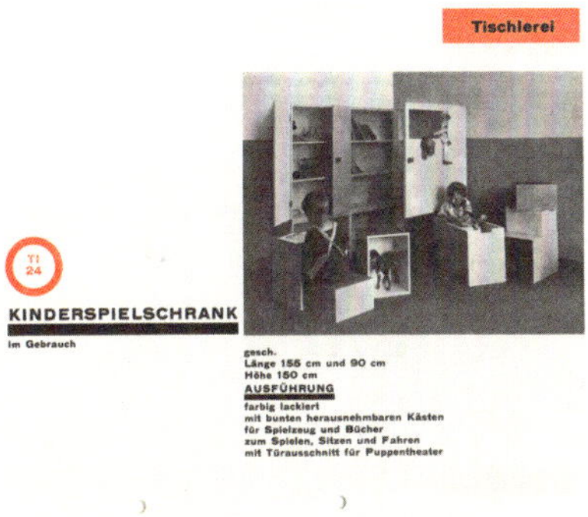
Tischlerei

TI 24

KINDERSPIELSCHRANK

im Gebrauch

gesch.
Länge 155 cm und 90 cm
Höhe 150 cm

AUSFÜHRUNG

farbig lackiert
mit bunten herausnehmbaren Kästen
für Spielzeug und Bücher
zum Spielen, Sitzen und Fahren
mit Türausschnitt für Puppentheater

Fig. 4. Herbert Bayer. Page from *Katalog der Muster* (Catalogue of Designs), sales catalogue of Bauhaus objects, layout by Bayer, showing children's toy closet (TI 24) designed by Alma Buscher in 1923. Dessau: Bauhaus, 1925. Letterpress. The Museum of Modern Art, New York. Jan Tschichold Collection, Gift of Philip Johnson.

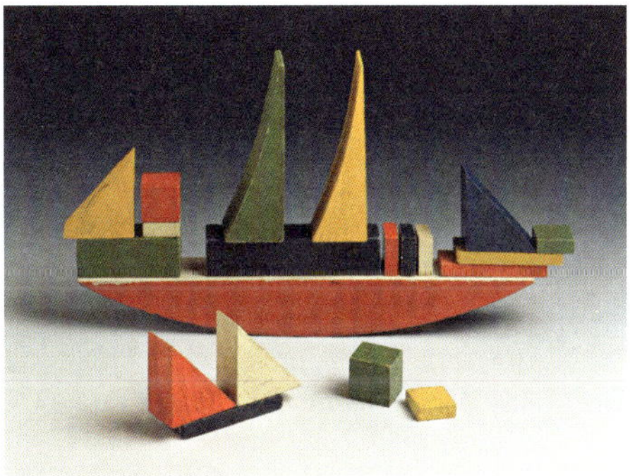

Fig. 5. Alma Buscher, "Ship" Building Toy, *Schiffsbauspiel*, 1923. 21 pieces, painted wood. Die Neue Sammlung—The International Design Museum Munich.

If the pleasures of Buscher's toy are to be found in the opportunities it provides for constructing several iterations of the ship itself, its challenges take place at the other end of playtime. Cut from two pieces of wood with no extraneous parts, it can be fit back into its box only with skill and concentration. This toy, Christine Mehring has written, "is about constructing and dismantling equally; in fact, dismantling becomes constructing and vice versa, for stashing the blocks away into their box is a puzzle that requires concentration and acumen."[10] The ship's modularity is thus revealed most clearly in reverse, in the effort expended in putting it away; the toy provides a certain set of simple tasks for younger children—build a ship with brightly colored blocks—while its storage arrangement demands the spatial skills of an older child, or perhaps an adult. The shift in scale from blocks to furniture to the house itself is thus conceptually intertwined with another shift in scale that takes place as the child inhabiting the Haus am Horn grows into adulthood.

At the Bauhaus, where architectural education emphasized playfulness, toys became as large as houses.[11] The Large-Scale Building Blocks designed by Gropius and Fred Forbat between 1921 and 1923 appeared, two years later, in Gropius's book *International Architecture*, where he captioned them "Models of Mass-Produced Houses. Variability of the same plan-type through the addition of repetitive spatial

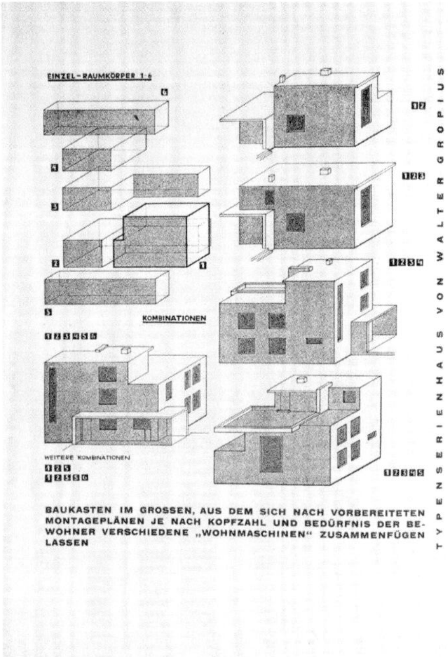

Fig. 6. Walter Gropius and Fred Forbat. "Large-Scale Building Blocks," 1921–23. From Alfred Meyer, ed., *Ein Versuchshaus des Bauhauses in Weimar* (Munich: Albert Langen, 1924), 8. Avery Architectural and Fine Arts Library, Columbia University.

units in different ways. Basic principles: integration of the greatest degree of repetition with the greatest possible variability" (fig. 6).[12] Here, the architectural building blocks appear infinitely interchangeable, their design examples labeled in descending order as a combination of numbered items. In presenting the complex beauty of fractal shapes, Mandelbrot hoped to demonstrate the sterility and banality of modernist architecture in an objective and scientific manner, yet the scale measurements—height, breadth, and depth—of the final designs are more perplexing, certainly, than those of the nineteenth-century Paris Opera House. Whatever constructions he hoped to invoke with the phrase "Bauhaus style, glass cube–type building," his discussions of scaling have placed these quasi-mythical modern creatures squarely in the camp from which he wished to exclude them: that of scaling objects.

For all its contradictory logic, Mandelbrot's endorsement of complexity in the form of "scaling" and his rejection of the modernist box stood firmly on the broader terrain of contemporary architectural culture. It followed by only fifteen years the publication of *Complexity and Contradiction in Architecture*, in which the architect Robert Venturi had countered Mies's famous declaration "less is more" with the assertion that "an architecture of complexity and contradiction . . . must embody the difficult unity of inclusion rather than the easy unity of exclusion. More is not less."[13] The Miesian modernist box had thus been under attack long before Mandelbrot dismissed it as "scalebound"; indeed, Venturi had noted that the Seagram building suffered not so much from a lack of complexity as from its refusal to express this structural complexity on the building's exterior. His fellow architect Louis Kahn, he noted, "once said that the Seagram building was like a beautiful lady with hidden corsets."[14] If Venturi lamented the invisibility of the building's wonderfully complex structural underclothing, Mandelbrot simply pronounced, with regard to its boxlike exterior, "it is fruitful to call Mies van der Rohe's buildings scalebound."[15]

—

Mandelbrot's curious essay had, as he revealed in a final footnote, first been "submitted to *Leonardo* in draft form in 1971, but its actual completion was delayed because I was working on different aspects of the subject." These efforts culminated in the publication in 1975 of *Les objets fractals*, the book he subsequently enlarged and modified for its English publication two years later as *Fractals: Form, Chance, and Dimension*.[16] One year after the book's appearance in English, as a note on his essay's first page attests, Mandelbrot submitted another iteration of the essay to *Leonardo*, and this is the version that appeared in print in 1981. Yet despite a decade of delays while he attended to related themes, he did not present "Scalebound or Scaling Shapes" as his final word on the subject; indeed, he explained, it is not even an essay at all. Rather, it was offered as a temporary placeholder, a "note"—albeit one accorded an initial capital letter (left uncorrected by the journal's copy editor) as if to reflect its higher status: "I regard this Note as a mere sketch, and I hope to expand the analysis when the opportunity arises."

Whether for lack of opportunity or lack of interest, Mandelbrot did not pursue the distinction between "scalebound and scaling shapes" with regard to the visual arts. What remains, then, is a doubly delayed effort—delayed for ten years after its first draft was submitted to *Leonardo* and for three years following its second submission—that presents itself as a rough sketch for further analysis to be carried out in the future. We might treat the essay as a retrospective indication of his earlier thinking: a sketch for his book on fractals, albeit printed after the book's publication. Such temporal slippage—a sketch for an essay plucked from the past, revised and resubmitted, standing in the place of promised future labor—may help explain the slipperiness of its arguments regarding "scalebound or scaling shapes," as well as their application to a broad range of only vaguely articulated forms. The temporal slippage involved in the essay's own publication serves another function, sliding the treatment of fractals backwards and forwards in time even as it expands to include natural creations and artistic works in all media. We might well consider Mandelbrot's essay a draft for future analyses, sketched out and never completed, but if its value as a historical document depends in part on the temporal direction of our gaze within Mandelbrot's own work, its significance depends as well on the level and intensity of our own interest in the topic of fractals.

1 Benoit Mandelbrot, "Scalebound or Scaling Shapes: A Useful Distinction in the Visual Arts and in the Natural Sciences," *Leonardo: Art, Science, Technology* 14 (1981): 45.
2 Ibid.
3 Ibid., 45 and 46.
4 Ibid., 45.
5 Ibid. In aligning architectural analysis with that of crystalline formations and Alpine visions, Mandelbrot followed a longstanding tradition, one of the primary referents for which was the nineteenth-century architect and theorist Eugène Emmanuel Viollet-le-Duc. See Jan von Brevern, *Blicke von Nirgendwo: Geologie in Bildern bei Ruskin, Viollet-le-Duc, und Civiale* (Munich: Wilhelm Fink, 2011), 93–121; on crystals as both symbols of abstraction and natural formations, see Regine Prange, *Das Kristalline als Kunstsymbol: Bruno Taut und Paul Klee* (New York: Georg Olms, 1991).
6 Mandelbrot, "Scalebound or Scaling Shapes," 45.
7 See Christopher Mead, *Charles Garnier's Paris Opera: Architectural Empathy and the Renaissance of French Classicism* (Cambridge, MA: MIT Press, 1991).
8 On the construction of and critical response to the Haus am Horn, see Wallis Miller, "Architecture, Building, and the Bauhaus," in *Bauhaus Culture: From Weimar to the Cold War*, ed. Kathleen James-Chakraborty (Minneapolis: University of Minnesota Press, 2006), 74–5.
9 For more on the toy, see Christine Mehring, "Alma Buscher, 'Ship' Building Toy, 1923," in *Bauhaus 1919–1933: Workshops for Modernity*, ed. Barry Bergdoll and Leah Dickerman, exh. cat. (New York: Museum of Modern Art exhibition catalog, 2009), 156–59.
10 Ibid., 159.
11 On playfulness at the Bauhaus, see Juliet Koss, "Bauhaus Theater of Human Dolls," *Modernism after Wagner* (Minneapolis: University of Minnesota Press, 2010), 207–43.
12 Gropius, quoted in Miller, "Architecture, Building, and the Bauhaus," 75.
13 Robert Venturi, *Complexity and Contradiction in Architecture* (New York: The Museum of Modern Art, 1966), 23.
14 Ibid., 56.
15 Mandelbrot's endorsement of architectural complexity has more recently proved useful for architects seeking theoretical inspiration for a shift away from the steel-and-glass modernist box but wishing to avoid the theoretical claims made for the postmodernism of previous generations; fractals and other late-twentieth-century developments in mathematics and physics have been central to the presentation of digital design. See, for example, Greg Lynn, *Folding in Architecture* (New York: Wiley-Academy, 2004); Lynn, *Folds, Bodies, and Blobs: Selected Essays* (Brussels: La Lettre Volée, 1998); Lynn, *Animate Form* (New York: Princeton Architectural Press, 1999); Ingeborg M. Rocker, "Calculus-Based Form: An Interview with Greg Lynn," *Architectural Design* 76 (2006): 88–95; and Jane Burry and Mark Burry, *The New Mathematics of Architecture* (London: Thames & Hudson, 2010). In this context, see also Paola Antonelli, interview with Benoit Mandelbrot, *Seed Magazine*, originally published March 24, 2008 [http://seedmagazine.com/content/article/paola_antonelli_benoit_mandelbrot/; accessed February 26, 2012].
16 Mandelbrot, "Scalebound or Scaling Shapes," 47.

The Eye Is Not Specialized!

Benoît Mandelbrot Talks with Nina Samuel

Nina Samuel (NS) Professor Mandelbrot, you have a very broad and interdisciplinary approach to working with scientific images. When reading the list of your publications between 1951 and 1975 one has to be astonished by the variety of your fields of study. How did you manage to switch so easily between completely different topics?

Benoît Mandelbrot (BM) I did not know, until I was about 19, that I have a powerful gift for shape. It might have been erased by regular schooling because students in mathematics are forced to do a great deal of calculations that end up by focusing all their energies. My schooling, on the contrary, was very much perturbed because of the German occupation of France. I was running around, trying to keep body and soul together. Therefore, I received a totally atypical education. In some ways you can say that I was under-educated; all I had were old illustrated books. Thus my visual abilities did not develop because of a school program but because of the circumstances.

At the age of 19 I discovered, to everybody's surprise, that every formula evoked in my mind a completely spontaneous shape. It was not a matter of willfully "visualizing" the formula, not a matter of choosing the best way to represent it. Once I heard a mathematical formula a shape came to my mind. Looking at the shape I was able in most cases to solve very difficult problems—instantly without hesitation. No question forced me to write down the usual two, five, or fifteen pages of mathematics. I just saw the solution right in front of my eyes.

The problem with calculations is that they become very specialized in the different fields. On the contrary, the eye is not specialized. It is a universal tool. So changing from one field to another was not an arduous task but an operation that even today I perform several times a day back and forth. I would work on something in the morning and if a visitor came to talk about something else, I would just switch to that something else. Throughout, I was mostly thinking in terms of pictures.

There are many reports of persons equally extreme in their visual dominance but very few choose to become scientists. Most scientists concentrate almost entirely on formal relations. Most visually skilled people go into architecture, design, house decorating, or something else, or become

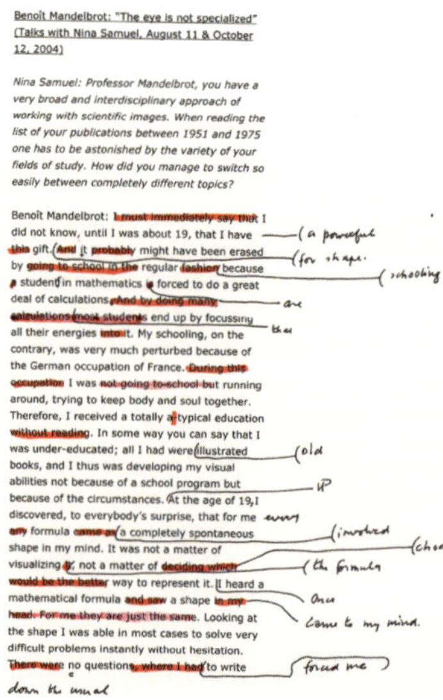
Benoît Mandelbrot: "The eye is not specialized" (Talks with Nina Samuel, August 11 & October 12, 2004)

Nina Samuel: Professor Mandelbrot, you have a very broad and interdisciplinary approach of working with scientific images. When reading the list of your publications between 1951 and 1975 one has to be astonished by the variety of your fields of study. How did you manage to switch so easily between completely different topics?

Benoît Mandelbrot: I must immediately say that I did not know, until I was about 19, that I have this gift. And it probably might have been erased by going to school in the regular fashion because a student in mathematics is forced to do a great deal of calculations. And by doing many calculations most students end up by focussing all their energies into it. My schooling, on the contrary, was very much perturbed because of the German occupation of France. During this occupation I was not going to school but running around, trying to keep body and soul together. Therefore, I received a totally atypical education without reading. In some way you can say that I was under-educated; all I had were illustrated books, and I thus was developing my visual abilities not because of a school program but because of the circumstances. At the age of 19, I discovered, to everybody's surprise, that for me any formula came as a completely spontaneous shape in my mind. It was not a matter of visualizing it, not a matter of deciding which would be the better way to represent it. I heard a mathematical formula and saw a shape in my head. For me they are just the same. Looking at the shape I was able in most cases to solve very difficult problems instantly without hesitation. There were no questions, where I had to write

Fig. 1. Benoît Mandelbrot. Handwritten amendments on printed manuscript page of this interview, 2005. Pen on paper. Private Collection.

artists, but I became a mathematician. Incidentally, I had very good grades in drawing. I was not a good artist at all, but I was extremely accurate.

NS By the time you studied in France pictures were not common in mathematical publications. Do you still notice the consequences of this "iconoclastic" period of Bourbaki today?[1]

BM I think Bourbaki is over. In fact, it has been over for some time, and I may have contributed to make them look foolish and boring. But the eye's revival will never be accepted by some survivors of that period. Younger people don't hesitate a minute when they can have a picture. They don't even think they are disobeying the great men of mathematics. They follow my advice because my advice is so obviously helpful. Maybe not everybody says so, but even some people of my generation who had never taken a hard look at a picture changed their minds after I showed them. And now, if they have something that can be visualized, they do not hesitate.

NS Looking at your publications before 1973, I didn't find any images other than graphs or diagrams. You once said you really became aware of the power of the eye around 1968, at the time you made your successful diagrammatic forgeries of the Nile and the stock market.[2] One could consider those graphs your first simulations of nature.

BM Yes, my first simulations of nature were implemented in the late sixties, when I worked on the river Nile. Many are reproduced in my Selecta book, titled *Gaussian Self-Affinity*.[3] They first appeared in the literature of hydrology. The reason for their lateness is simple: Before the sixties, there were no computer graphics. This was putting me in a very clumsy situation. I saw pictures in my head, but could not draw them because of the lack of tools. The only machine that existed was very clunky, done by a company called Calcomp. It was used to design machinery, not to simulate nature. My initial goal was very simple. I wanted to share with everybody the mental pictures that I had in my head and that nobody else could see. But I immediately realized by seeing the actual pictures that they also helped me think of other pictures and helped me in other ways. Therefore, the pictures began by being purely pedagogical. But they very rapidly moved to being absolutely fundamental for my own thinking.

NS So the pictures you produced by means of computer graphics were really the pictures you had in mind? Was there a correspondence?

BM Well, the pictures that I found were not exactly those I had in my mind but very close.

NS Could you describe the difference between what you had in mind and what came out of the computer?

BM The simulations were invariably much more detailed, much more involved. It became fortunate that I was working at IBM at that time.[4] This was not because IBM had machines to do what I wanted—IBM had nothing, nobody had machines. More precisely, I think that Los Alamos had machines, but nobody had access to them. Besides, those machines were not used for this purpose at all. The advantage of being at IBM was that several friends were skilled engineers or programmers. I succeeded in convincing them that my kind of simulation was going to be amusing and very avant-garde frontier work. I motivated them to do work they had never imagined, to write software where no software existed and had never been envisioned. We had to go around, scrounge whatever was available, and transform it into machines to make visible what I had in my mind. Soon, disregarding the initial motivation, the computer and the pictures became integrated parts of my thought process. By changing the inputs, I could follow how the pictures changed, and I made many discoveries.

NS What I find very interesting is that the fractal theory started as a description but turned out to be an explanation for many natural phenomena. Concerning the images, where and when would you say that the description became explanation?

BM This is a very delicate question, because, slowly and gradually, I realized

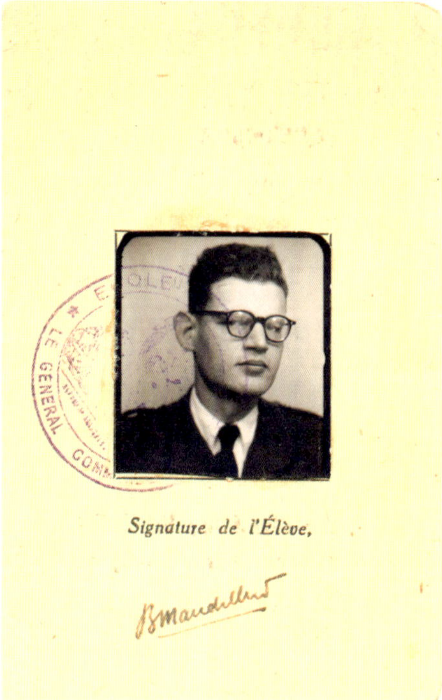

Fig. 2. Benoît Mandelbrot's École polytechnique identity card, ca.1946/47. © Collections École polytechnique. Gift of Madame Aliette Mandelbrot.

that what I was studying was not just a collection of separate phenomena but a general topic, called "roughness." Roughness is one of those basic sensations of man for which there was no theory; there were theories for noisiness, loudness, brightness, heaviness, hotness—but not for roughness. I started on the way to a theory by identifying the first-ever intrinsic measure of roughness. That step was essential and universal, but still a matter of description. The passage from description to theory is completely different. It depends on the field you are in. In some cases the passage was complete. In other cases, there is no passage: Fractals only yielded a description and nothing else. But let it be: To have an exact description with a number for roughness is much more valuable than to rush to a bad explanation.

NS Let's change to your discovery of the Mandelbrot set: You actually saw the set for the first time in 1979, but you didn't identify it until 1980. Do you remember and could you define the very moment when seeing switched into discovering? When and why did you decide to look more closely at those confusing "messy" parts of your visualizations?

BM I could not give an exact date but the fact that I am extremely meticulous is a matter of personality. I was a visitor at Harvard at that time, without my own equipment, only equipment many people used almost twenty-four hours a day. The screen was very bad and I could not hope for clean pictures. So I kept looking at the very dirty early pictures of that set again and again and again. Soon, I observed that two pieces of dirt were symmetrical. This fact made me very, very concerned. I drove to IBM and arranged for what I thought would be clean pictures to be printed. But the pictures were even dirtier, because a better printer showed far more interesting detail. Then I became seriously excited. These big discoveries rewarded me for respecting the pictures by letting the brain interact with them. This procedure has been successful many, many times. My assistant once brought a file and said: "I just thought you may like these two pictures. They look identical and you may want to have them as a demo for your lectures." I looked at the pictures and instantly saw the difference. He insisted, "they are identical, I see no difference." But the difference I had seen turned out to be very significant.

NS So you used the method of visual comparison.

BM Well, I automatically look for symmetries, identities, differences, which is an elementary way of handling pictures. Everybody who is accustomed to pictures knows that.

NS Tragically, even in your first publication of the whole Mandelbrot set in 1980 the publisher left out those "dirt-speckles," which were the most important parts.

BM Yes, it was incredible! But it was less a tragedy than a farce.

NS Do you think the scientific world was afraid of the "mathematical dirt" that might have soiled the image of high technology as a symbol of precision?

BM Publishing is a business in which many people handle each document. I imagine that the first person read my instructions and printed the dirt. But the person who made the last visual check must have cleaned off these dots. Luckily, they are discussed in the paper! The first main point of the paper was my discovery of dots! After they were suppressed I had to add them in each reprint by hand. This became one of the big jokes of the time. I would say that this episode confirms something very familiar: What you don't expect to see, you don't see. One must learn to see.

NS I would like to retrace how you made those pictures part of your thinking process. It was in April 1979 when you started with experiments of non-quadratic maps, which preceded the discovery of the Mandelbrot set.

BM To be accurate, the first picture I thought worth reproducing dates from April 1979. There were many older pictures but I found them less interesting, so there is a very large number of unpublished ones (cf. Found Images I, cats. 21–39). When one begins by exploring a completely unknown territory, like a new mountain that nobody has scaled before, I think it is very important to first walk around and get a feeling of what's happening. So I made many very preliminary pictures that I did not publish, because they would have overwhelmed the book.[5]

NS When and how did the shapes begin to get interesting, and when did you realize that there was something there you wanted to pursue?

BM All were very interesting, but they were increasingly hard to interpret. It is always best to open a new direction in science with a simple case. But I did not suspect—nobody did—how complicated the study of $z^2 + c$ was going to be. So instead I first studied a ratio of higher-order polynomials—generalizing a ratio known to be interesting. But it turned out to be too complicated for further study. As usual, I did hundreds of pictures, the same way a good photographer needs hundreds of pictures to catch the right glimmer in the eye of the subject or the right light on a landscape. In fact, I had to convince each new assistant that it was much better to make too many pictures than too few. Some we didn't even print because they were uninteresting. It seemed much better to think for a long time before starting a new picture, but in fact the situation at IBM was very unusual. At night or on weekends, computers that were very powerful for that time cost nothing. On the other hand, assistants were very well paid. So I had few and kept telling them: "I pay you so much that if you must spend one hour thinking whether it is better to proceed this or that way, do it both ways, this will take you less time." So my study of that complicated formula involved many, many pictures but they were very difficult to interpret.

NS But how did you go on with these forms? Did you classify them by shapes?

BM No, I tried classifying what I had

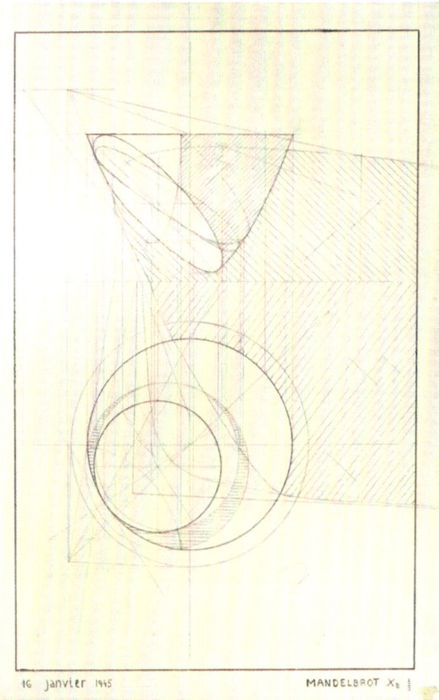

Fig. 3. Benoît Mandelbrot. Geometry exercise from a special mathematics class, January 16, 1945. © Collections École polytechnique. Gift of Madame Aliette Mandelbrot.

but achieved nothing and gave up. Then I moved to the simplest map, which was $z^2 + c$. That map is simple enough so that I could get the Mandelbrot set easily and in great detail. Depending on the value of parameter c, the Julia set corresponding to was either connected or disconnected—that is the distinction that defines the Mandelbrot set. And depending where c is located with respect to the Mandelbrot set, the corresponding Julia set can take all kinds of shapes: one arm, or two, or fifteen. So I conceived the Mandelbrot set as a notion of a map for the various shapes of the Julia sets. Of course the term "Mandelbrot set" came much later. I was calling it a "map," the μ-map because c was called μ, or the λ-map, because the quadratic polynomial has an alternative reduced form, namely $\lambda z (1-z)$. Depending upon where you are in this map, the shape of the Julia set was different.

NS That means that on the one side you had the picture of the Mandelbrot set and on the other side the pictures of the Julia sets and then you searched for a resemblance between the two?

BM Things were less organized and therefore more complicated. Early on, we didn't label the pictures because when one begins, one is neglectful. They were accumulating on my desk and at one point I could no longer remember which was which. So I did two things: In a police action, I asked my assistant to label each picture, but in addition I started thinking. The fact that I could confuse the two kinds of pictures was an irritating but very interesting property. It led me to conjecture that, in some sense to be specified, the ends of the arms of the Julia sets and of the Mandelbrot set had the same shape. Again, it was not yet called a Mandelbrot set but a μ-map or a λ-map.

NS So you made the discovery in the moment when you realized that the forms were interchangeable between each other, that they belong very close together...

BM Well, very close together but in a very complicated fashion. The Julia set is in a certain sense homogenous. Corners of the boundary of the filled-in Julia set have the same shape in every point, except for deformations. In contrast to this, the Mandelbrot

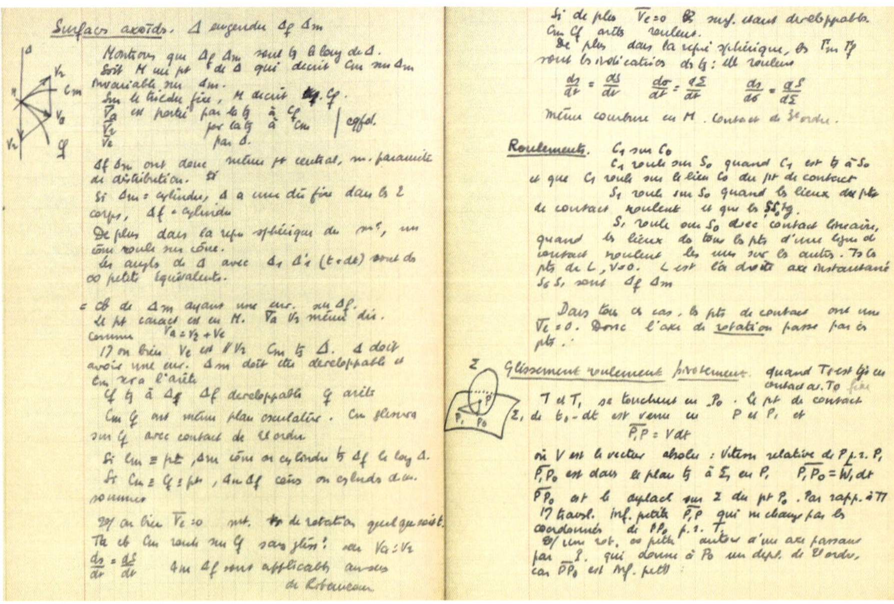

Fig. 4. Benoît Mandelbrot. Notes taken during the geometry class of Gaston Julia, 1945–47. Ink on paper. © Collections École polytechnique. Gift of Madame Aliette Mandelbrot.

set includes in itself at different corners little images of every Julia set that exists. For each value of c you have a single Julia set, but close to that value of *c* the Mandelbrot set's boundary looks very much like the Julia set that it produces.

NS So it is only analogous, not exactly the same.

BM Not the same. The exact proof took five years. It was made by the Chinese mathematician Tan Lei. My statement involved a certain amount of looseness. She made it correct and proved it. It was a very, very difficult proof.

NS Did she prove it also by pictures or only through mathematics?

BM Only through mathematics. A proof can only proceed through mathematics. Pictures are suggestive and can lead to conjectures. A picture by itself does not prove anything.

NS Word goes out that today there is a kind of pictorial turn in the natural sciences. How do you evaluate that?

BM In regard to this, there is a big difference between mathematics and science. Science has always had a pictorial element. There have always been hand-drawn pictures of birds, ants, or the eddies formed when water flows around an obstacle. Then there were photographs. Mathematics has had a different history, with the disappearance of pictures. So in a certain sense, I brought mathematics back into the norm.

NS Professor Mandelbrot, thank you very much.

This interview was first published as "'The eye is not specialized!' Benoît Mandelbrot talks with Nina Samuel," *Bildwelten des Wissens. Kunsthistorisches Jahrbuch für Bildkritik*, no. 3 (2005): Digitale Form, 32–39.

1 For an explanation of Bourbaki, please see Nina Samuel, "The Visibility of Islands," in this volume, 24–28.

2 "Forgery" is a term used by both Mandelbrot and programmer Richard Voss for fractal simulations, see Benoît Mandelbrot, "Recorded History and Personal Recollections," in *Gaussian Self-Affinity and Fractals* (New York: Springer, 2002), especially 222; and Richard F. Voss, "Random Fractal Forgeries: From Mountains to Music," in *Science and Uncertainty*, ed. Steve Nash (Northwood, England: Science Reviews, 1985), 68–85.

3 Mandelbrot, *Gaussian Self-Affinity and Fractals*.

4 Benoît Mandelbrot worked for IBM since 1958 (from 1958 to 1974 as a research staff member; from 1974 to 93 as an IBM Fellow, and from 1993 until his death as an IBM Fellow Emeritus).

5 Benoît Mandelbrot, *Fractals and Chaos: The Mandelbrot Set and Beyond* (New York: Springer, 2004).

Bibliography

Abraham, Ralph, and Yoshisuke Ueda, eds., *The Chaos Avant-Garde: Memories of the Early Days of Chaos Theory*. Singapore: World Scientific, 2000.

Adriani, Götz, Winfried Konnetz, and Karin Thomas. *Joseph Beuys*. Cologne: Dumont Buchverlag, 1994.

Antonelli, Paola. "Interview with Benoit Mandelbrot." *Seed Magazine* (March 24, 2008). Accessed February 26, 2012, http://seedmagazine.com/content/article/paola_antonelli_benoit_mandelbrot.

Arnheim, Rudolf. *Art and Visual Perception: A Psychology of the Creative Eye*. Berkeley: University of California Press, 1954.

——. *Visual Thinking*. Berkeley: University of California Press, 1969.

——. *Entropy and Art: An Essay on Disorder and Order*. Berkeley: University of California Press, 1971.

——. "Perceiving and Portraying." Review of *The Image and the Eye: Further Studies in the Psychology of Pictorial Representation*, by E. H. Gombrich. *Times Literary Supplement*, October 29, 1982.

——. "Symmetry and the Organization of Form: A Review Article." *Leonardo* 21 (1988): 273–76.

——. "From Chaos to Wholeness." *The Journal of Aesthetics and Art Criticism* 54 (1996): 117–20.

Aubin, David, and Amy Dahan Dalmedico. "Writing the History of Dynamical Systems and Chaos: Longue Durée and Revolution, Disciplines and Cultures." *Historia Mathematica* 29 (2002): 271–339.

Bailer-Jones, Daniela M. *Scientific Models in Philosophy of Science*. Pittsburgh: University of Pittsburgh Press, 2009.

Baillaud, Benjamin, and Henry Bourget, eds. *Correspondance d'Hermite et de Stieltjes*. Paris: Gauthier-Villars, 1905.

Barcellos, A. "Interview of B. B. Mandelbrot." In *Mathematical People*. Edited by Donald J. Albers and Gerald L. Alexanderson, 205–25. Boston: Birkhäuser, 1984.

Barnsley, Michael F. *Fractals Everywhere*. Boston: Academic Press, 1988.

Bätschmann, Oskar. *Entfernung der Natur. Landschaftsmalerei 1750–1920*. Cologne: DuMont, 1989.

Baudrillard, Jean. *Die Agonie des Realen*. Berlin: Merve, 1978.

——. *Subjekt und Objekt: fraktal. Vortrag im Kunstmuseum Bern*. Bern: Wabern, 1986.

Belting, Hans. *Bild und Kult: eine Geschichte des Bildes vor dem Zeitalter der Kunst*. Munich: Beck, 2000.

Berger, J. M., and Benoît Mandelbrot. "A New Model for Error Clustering in Telephone Circuits." *IBM Journal of Research and Development* 7 (1963): 224–36.

Bernoulli, Jacob. *Wahrscheinlichkeitsrechnung (Ars conjectandi)*. Translated by Robert Haussner. Frankfurt am Main: Thun, 1999.

Beuys, Joseph. *Zeichnungen 1947–59*. Cologne: Schirmer Verlag, 1972.

Biderman, A. D. "The Playfair Enigma: The Development of the Schematic Representation of Statistics." *Information Design* 6 (1990): 3–25.

Bigg, Charlotte. "A Visual History of Jean Perrin's Brownian Motion Curves." In *Histories of Scientific Observation*. Edited by Lorraine Daston and Elizabeth Lunbeck, 156–79. Chicago: University of Chicago Press, 2011.

Boehm, Gottfried. "Die Wiederkehr der Bilder." In *Was ist ein Bild?* Edited by Gottfried Boehm, 11–38. Munich: Fink, 1994.

——. "Zwischen Auge und Hand. Bilder als Instrumente der Erkenntnis." In *Mit dem Auge denken. Strategien der Sichtbarmachung in wissenschaftlichen und virtuellen Welten*. Edited by Bettina Heintz and Jörg Huber, 43–54. Vienna: Springer, 2001.

——. "The Form of the Formless: Abstract Expressionism and Art Informel." In *Action Painting*. Edited by Delia Ciuha, 38–46. Ostfildern, Germany: Hatje Cantz, 2008.

Bonnefoit, Régine. *Die Linientheorien von Paul Klee*. Petersberg: Imhof, 2009.

Borel, Armand. *Mathematik: Kunst und Wissenschaft*. Munich: Die Stiftung, 1982.

Bourbaki, Nicolas. *Topologie générale*, book 3 of *Élements de Mathématique*. Paris: Hermann, 1958.

——. "Die Architektur der

Mathematik." In *Mathematiker über Mathematik*. Edited by Michael Otte, 140–59. Berlin: Springer, 1974.

Bowler, Richard C. "Mediating Creative Nature and Human Needs in Early German Political Economy." *History of Political Economy* 40 (2008): 633–69.

Bozzetto, R. M. P., and Russell Taylor. "Moreau's Tragi-Farcical Island." *Science Fiction Studies* 20 (1993): 34–44.

Bredekamp, Horst. "Mimesis, grundlos." *Kunstforum International* 114 (1991): 278–88.

———. "Der simulierte Benjamin. Mittelalterliche Bemerkungen zu seiner Aktualität." In *Frankfurter Schule und Kunstgeschichte*, 117–40. Berlin: Reimer, 1992.

———. *The Lure of Antiquity and the Cult of the Machine: The Kunstkammer and the Evolution of Nature, Art, and Technology*. Princeton: Markus Wiener, 1995.

———. "Das Bild als Leitbild. Gedanken zur Überwindung des Anikonismus." In *LogIcons: Bilder zwischen Theorie und Anschauung*. Edited by Ute Hoffmann, Bernward Joerges, and Ingrid Severin, 225–47. Berlin: Sigma, 1997.

———. "Die Unüberschreitbarkeit der Schlangenlinie." In *Minimal Concept: zeichenhafte Sprachen im Raum*. Edited by Christian Schneegass, 205–208. Dresden: Verlag der Kunst; Berlin: Akademie der Künste, 2001.

———. "A Neglected Tradition? Art History as 'Bildwissenschaft.'" *Critical Inquiry* 29 (2003): 418–28.

———. "Die Zeichnende Denkkraft. Überlegungen zur Bildkunst der Naturwissenschaften." In *Einbildungen*. Edited by Jörg Huber, 155–71. Zurich: Springer, 2005.

———. "John Cage and the Principle of Chance." In *Music and the Aesthetics of Modernity*. Edited by Karol Berger and Anthony Newcomb, 99–107. Cambridge, MA: Harvard University Press, 2005.

———. "Der Mensch als 'zweiter Gott.' Motive der Wiederkehr eines kunsttheoretischen Topos im Zeitalter der Bildsimulation." In *Bilder bewegen. Von der Kunstkammer zum Endspiel*. Edited by Jörg Probst, 106–20. Berlin: Wagenbach, 2007.

———. *Theorie des Bildakts: Über das Lebensrecht des Bildes*. Frankfurt am Main: Suhrkamp, 2010.

Bredekamp, Horst, Angela Fischel, Birgit Schneider, and Gabriele Werner. "Bildwelten des Wissens." In *Bilder in Prozessen. Bildwelten des Wissens. Kunsthistorisches Jahrbuch für Bildkritik* 1 (2003): 9–20.

Bredekamp, Horst, Birgit Schneider, and Vera Dünkel. *Das Technische Bild. Kompendium zu einer Stilgeschichte wissenschaftlicher Bilder*. Berlin: Akademie, 2008.

Brooks, Robert, and J. Peter Matelski. "The Dynamics of 2-generator Subgrounds of PSL (2,C)." In *Riemann Surfaces and Related Topics: Proceedings of the 1978 Stony Brook Conference, Annals of Mathematics Studies* 97. Edited by Irwin Kra and Bernard Maski, 65–71. Princeton: Princeton University Press, 1981.

Bruhn, Matthias, and Vera Dünkel. "The Image as Cultural Technology." In *Visual Literacy*. Edited by James Elkins, 165–78. New York and London: Routledge, 2008.

Buchanan, Mark. "The Irregularity of Reality." *Nature Physics* 7 (2011): 184.

Bullynck, M. "Markovketten—Variationen im Computer." In *Andrej A. Markov. Berechenbare Künste. Mathematik, Poesie, Moderne*. Edited by Ph. von Hilgers and Wladimir Velminski, 158–73. Zurich and Berlin: diaphanes, 2007.

Burnet, Thomas. *The Sacred Theory of the Earth, Containing an Account of the Original of the Earth*. London, 1681.

Burry, Jane and Mark Burry. *The New Mathematics of Architecture*. London: Thames & Hudson, 2010.

Busch, Werner. *Das sentimentalische Bild: die Krise der Kunst im 18. Jahrhundert und die Geburt der Moderne*. Munich: Beck, 1993.

Cannon, James W. "Review of *The Fractal Geometry of Nature* by Benoît B. Mandelbrot." *The American Mathematical Monthly* 91 (1984): 594–98.

Chabert, Jean-Luc. "Un demi-siècle de fractales: 1870–1920." *Historia Mathematica* 17 (1990): 339–65.

Chandler, Alfred D. Jr. *Strategy and Structure: Chapters in the History of the American Industrial Enterprise*. Cambridge, MA: MIT Press, 1962.

Changeux, Jean-Pierre, and Alain Connes. *Conversations on Mind, Matter, and Mathematics*. Princeton: Princeton University Press,1995.

Cheung, Tobias. "From the Organism of a Body to the Body of an Organism: Occurrence and Meaning of the Word 'Organism' from the Seventeenth to the Nineteenth Centuries." *The British Journal for the History of Science* 39 (2006): 319–39.

———. ed. *Transitions and Borders Between Animals, Humans, and Machines, 1600–1800*. Leiden, The Netherlands: Brill, 2010.

Clausberg, Karl. "Feigenbaum und Mandelbrot. Neue Symmetrien zwischen Kunst und Naturwissenschaften." *Kunstforum International* 85 (1986): 86–93.

———. "Am Ende Kunstgeschichte? Künstliche Wirklichkeiten aus dem Computer." In *Kunstgeschichte—aber wie? Zehn Themen und Beispiele*. Edited by Clemens Fruh, Raphael Rosenberg, and Hans-Peter Rosinski, 259–93. Berlin: Reimer, 1989.

Crary, Jonathan. *Techniques of the Observer: On Vision and Modernity in the Nineteenth Century*. Cambridge, MA: MIT Press, 1990.

Crystal, David, ed., *The Cambridge Encyclopedia of Language*. Cambridge: Cambridge University Press, 1987.

Culling, W. E. H. "A New View of the Landscape." *Transactions of the Institute of British Geographers* 13 (1988): 345–60.

Dackerman, Susan, ed. *Prints and the Pursuit of Knowledge in Early Modern Europe*. Cambridge, MA: Harvard Art Museums, 2011.

Daston, Lorraine, and Katherine Park. *Wonders and the Order of Nature, 1150–1750*. New York: Zone Books, 1998.

Daston, Lorraine, and Michael Stolleis, eds. *Natural Law and Laws of Nature in Early Modern Europe: Jurisprudence, Theology, Moral, and Natural Philosophy*. Farnham, UK: Ashgate, 2008.

Davis, Monte. "Profile of Benoît B. Mandelbrot." *Omni Magazine*. February 5, 1984.

De Beaumont, Léonce Élie. *Notice sur les systèmes des montagnes*. Paris: P. Bertrand, 1852.

Diedrichs, Christof L. *Vom Glauben zum Sehen: Die Sichtbarkeit der Reliquie im Reliquiar: Ein Beitrag zur Geschichte des Sehens*. Berlin: Weissensee, 2001.

Dieudonné, Jean A. "The Work of Nicholas Bourbaki." *The American Mathematical Monthly* 77 (1970): 134–45.

Dieudonné, Jean A., Françios Guénard, and Gilbert Lelièvre, eds. *Penser les mathématiques*. Paris: Seuil, 1982.

Ditzen, Stefan. "Der Satyr auf dem Larvenrücken. Zum Verhältnis von instrumentellem Sehen und Bildtraditionenen." In *Konstruierte Sichtbarkeiten. Wissenschafts- und Technikbilder seit der Frühen Neuzeit*. Edited by Martina Hessler, 41–56. Munich: W. Fink, 2006.

Douady, Adrien. "Systèmes dynamiques holomorphes." *Séminaire N. Bourbaki* 599 (1982): 39–63.

Douady, Adrien, and John Hamal Hubbard. "Itération des polynômes quadratiques complexes." *Comptes Rendus* (Paris) 294 (1982): 123–26.

Du Bois-Reymond, Paul. "Versuch einer Classification der willkürlichen Functionen reeller Argumente nach ihren Aenderungen in den kleinsten Intervallen." *Journal für die reine und angewandte Mathematik* 79 (1875): 21–37.

Dyson, Freeman. "Characterizing Irregularity." A review of *Fractals: Form, Chance and Dimension*, by Benoît Mandelbrot. *Science* 200 (1978): 677–78.

Einstein, Albert. "Geometrie und Erfahrung." *Sitzungsberichte der Preussischen Akademie der Wissenschaften* (January 1921): 123–30.

Fatou, Pierr. "Sur les équations fonctionelles." *Bulletin de la Soc. Math. Fr.* 47 (1919): 161–271.

Feller, William. *An Introduction to Probability Theory and Its Applications*. New York: Wiley, 1957.

Fischel, Angela. *Natur im Bild: Zeichnung und Naturerkenntnis bei Conrad Gessner und Ulisse Aldrovandi*. Berlin: Mann, 2009.

Fischer, Gerd, ed. *Mathematische Modelle. Aus den Sammlungen von Universitäten und Museen*. Berlin: Akademie-Verlag, 1986.

Fludd, Robert. *The Origin and Structure of the Cosmos*. Translated by Patricia Tahil. Edinburgh: Magnum Opus Hermetic Sourceworks, 1982.

Flusser, Vilém. "Curie's Children: Vilém Flusser on an Unspeakable Future." *Artforum* 28 (1990): 22–23.

Fournier, Alain, Don Fussell, and Loren Carpenter. "Computer Rendering of Stochastic Models." *Communications of the ACM 25* (1982): 371–84.

Frey, Pierra A., and Lise Grenier, eds. *Viollet-le-Duc et la Montagne*. Grenoble, France: Glénat, 1993.

Frigg, Roman, and Stephen Hartmann. "Models in Science." In *The Stanford Encyclopedia of Philosophy*, February 27, 2006. Accessed May 21, 2012, http://plato.stanford.edu/archives/sum2009/entries/models-science.

Galison, Peter. "The Suppressed Drawing: Paul Dirac's Hidden Geometry." *Representations* 72 (2000): 145–66.

———. "Image of Self." In *Things That Talk: Object Lessons from Art and Science*. Edited by Lorraine Daston, 257–94. New York: Zone Books, 2004.

Geimer, Peter, ed. *Ordnungen der Sichtbarkeit. Fotografie in Wissenschaft, Kunst und Technologie*. Frankfurt am Main: Suhrkamp, 2002.

Geimer, Peter. *Bilder aus Versehen: eine Geschichte fotografischer Erscheinungen*. Hamburg: Philo Fine Arts, 2010.

———. "Zur zeichnerischen Simulation von Natur und natürlicher Lebendigkeit." *Zeitschrift für Ästhetik und Allgemeine Kunstwissenschaft* 34 (1989): 243–79.

Gigerenzer, Gerd. *The Empire of Chance: How Probability Changed Science and Everyday Life*. Cambridge: Cambridge University Press, 1990.

Glaser, Peter. "Das Kolumbus-Gefühl. Entdeckungen in einer virtuellen Welt." In *Computerkultur oder The Beauty of Bit & Byte*. Edited by Michael Weisser, 19–40. Bremen, Germany: TMS, 1989.

Gleick, James. *Chaos: Making a New Science*. New York: Viking, 1987.

Gludovatz, Karin. "Malerische Worte. Die Künstlersignatur als Schrift-Bild." In *Schrift: Kulturtechnik zwischen Auge, Hand und Maschine*. Edited by Gernot Grube, Werner Kogge, and Sybille Krämer, 313–28. Munich: W. Fink, 2005.

Gombrich, Ernst H. *Art and Illusion: A Study in the Psychology of Pictorial Representation*. Princeton, NJ: Bollingen Series, 1956.

——. *The Sense of Order: A Study in the Psychology of Decorative Art.* Oxford: Phaidon, 1979.

Gould, Stephen Jay. *Questioning the Millennium: A Rationalist's Guide to a Precisely Arbitrary Countdown.* London: Vintage, 1998.

Gowans, Alan. *The Unchanging Arts: New Forms for the Traditional Functions of Art in Society.* Philadelphia: Lippincott, 1971.

Greene, Mott T. *Geology in the Nineteenth Century: Changing Views of a Changing World.* Ithaca, NY: Cornell University Press, 1982.

Groh, Dieter, and Ruth Groh. *Weltbild und Naturaneignung. Zur Kulturgeschichte der Natur.* Frankfurt am Main: Suhrkamp, 1991.

Grossklaus, Götz, and Ernst Oldemeyer, eds. *Natur als Gegenwelt. Beiträge zur Kulturgeschichte der Natur.* Karlsruhe, Germany: Loeper, 1983.

Grote, Andreas, ed. *Macrocosmos in Microcosmo: Die Welt in der Stube. Zur Geschichte des Sammelns 1450 bis 1800.* Opladen, Germany: Leske + Budrich, 1994.

Gugerli, David, and Barbara Orland. *Ganz normale Bilder. Historische Beiträge zur visuellen Herstellung von Selbstverständlichkeit.* Zurich: Chronos, 2002.

Hadamard, Jacques. *An Essay on the Psychology of Invention in the Mathematical Field.* New York: Dover, 1954.

Halmos, Paul R. "Nicolas Bourbaki." *Scientific American* 196 (1957): 88–99.

Hamilton, Richard. "Review of *The Visual Arts Today*, edited by György Kepes." *Design* 154 (1961): 91.

Hampe, Michael. *Eine kleine Geschichte des Naturgesetzbegriffs.* Frankfurt am Main: Suhrkamp, 2007.

Handelman, Sigmund. "A High-Resolution Computer Graphics System." *IBM Systems Journal* 19 (1980): 356–66.

Hardy, G. H. "A Mathematician's Apology." In *The World of Mathematics*, Vol. 4. Edited by James Roy Newman, 2027–38. New York: Simon and Schuster, 1956.

Hargittai, Magdolna, and István Hargittai. *Candid Science Four.* London: Imperial College Press, 2004.

Haskell, Francis. *History and Its Images: Art and the Interpretation of the Past.* New Haven: Yale University Press, 1993.

Hausdorff, Felix. "Dimension und äusseres Mass." *Mathematische Annalen* 79 (1919): 157–79.

Heck, Karsten. "Der Schreibtisch als Denkraum. Heinrich von Geymüllers 'Tableau graphique.'" In *Heinrich von Geymüller (1839–1909).* Edited by Josef Ploder and Georg Germann, 54–63. Basel: Reinhardt, 2009.

Heims, Steve J. *Constructing a Social Science for Postwar America: The Cybernetics Group, 1946–1953.* Cambridge: Cambridge University Press, 1993.

Heintz, Bettina. *Die Herrschaft der Regel. Zur Grundlagengeschichte des Computers.* Frankfurt am Main: Campus Verlag, 1993.

——. "Zeichen, die Bilder schaffen." In *Bilder—Symbole—Metaphern, Visualisierung und Informierung in der Moderne.* Edited by Johanna Hofbauer, Gerald Prabitz, and Josef Wallmannsberger, 47–77. Vienna: Passagen, 1995.

Heintz, Bettina, Arnold Benz, and Jörg Huber, eds. *Mit dem Auge denken: Strategien der Sichtbarmachung in wissenschaftlichen und virtuellen Welten.* Zurich: Voldemeer, 2001.

Hennig, Jochen. "Bildtradition und Differenz. Visuelle Erkenntnisgewinnung in der Wissenschaft am Beispiel der Rastertunnelmikroskopie." In *Das Technische Bild. Kompendium zu einer Stilgeschichte wissenschaftlicher Bilder.* Edited by Horst Bredekamp, Birgit Schneider, and Vera Dünkel, 86–95. Berlin: Akademie Verklag, 2008.

——. *Bildpraxis: Visuelle Strategien in der frühen Nanotechnologie* Bielefeld: Transcript, 2011.

Hessler, Matina, ed. *Konstruierte Sichtbarkeiten. Wissenschafts- und Technikbilder seit der Frühen Neuzeit.* Munich: W. Fink, 2006.

Hilbert, David. "Über die stetige Abbildung einer Linie auf ein Flächenstück." *Mathematische Annalen* 38 (1891): 459–60.

——. *Grundlagen der Geometrie. Festschrift zur Feier der Enthüllung des Gauss-Weber-Denkmals in Göttingen.* Leipzig: B. G. Teuber, 1899.

——. "Mathematische Probleme." In *Die Hilbertschen Probleme.* Edited by P. S. Alexandrov, 22–80. Leipzig: Akademische Verlagsges, Geest & Portig, 1983.

von Hilgers, Ph., and Wladimir Velminski, eds. *Andrej A. Markov. Berechenbare Künste. Mathematik, Poesie, Moderne.* Zurich: diaphanes, 2007.

Hill, Edward. *The Language of Drawing.* Englewood Cliffs, NJ: Prentice-Hall, 1966.

Hogarth, William S. *Analyse der Schönheit.* Translated by Jörg Heininger. Hamburg: Philo Fine Arts, 2008.

Holeczek, Bernhard, and Lida von Mengden, eds. *Zufall als Prinzip: Spielwelt, Methode und System in der Kunst des 20. Jahrhunderts.* Exh. cat. Ludwigshafen am Rhein, Wilhelm-Hack-Museum, 1992.

Holzhey, Magdalena. *Im Labor des Zeichners: Joseph Beuys und die Naturwissenschaft.* Berlin: Reimer, 2009.

Hopwood, Nick, Simon Schaffer, and Jim Secord, eds. *Seriality and Scientific Objects in the Nineteenth Century*, 251–494. Cambridge: Science History Publishing, 2010.

Hoyle, Fred. *Astronomy and Cosmology: A Modern Course.* San Francisco: W. H. Freeman & Co, 1975.

Hubbard, John Hamal. Preface to *The Mandelbrot Set: Theme and Variations*. Edited by Tan Lei, xiii–xx. Cambridge: Cambridge University Press, 2000.

Huschke, Ralph E., ed. *Glossary of Meteorology*. Boston: American Meteorological Society, 1959.

Imdahl, Max. "Jacob van Ruisdaels 'Die Mühle von Wijk.'" In *Max Imdahl: Gesammelte Schriften*. Edited by Gundolf Winter, 345–84. Frankfurt am Main: Suhrkamp, 1996.

Jackson, Allyn. "Comme Appelé du Néant—As If Summoned from the Void: The Life of Alexandre Grothendieck, II." *Notices of the AMS* 51 (2004): 1196–212.

Jacobi, Daniel, and Bernard Schiele. "Scientific Imagery and Popularized Imagery: Differences and Similarities in the Photographic Portraits of Scientists." *Social Studies of Science* 19 (1989): 731–53.

Janecke, Christian. *Kunst und Zufall. Analyse und Bedeutung.* Nuremberg: Verlag für Moderne Kunst, 1995.

Jones, Caroline A., and Peter Galison, eds. *Picturing Science, Producing Art.* New York: Routledge, 1998.

Julia, Gaston. "Mémoire sur l'itération des fonctions rationelles." *Journal de Math. pures et appliquées* 8.1 (1918): 47–245.

Jürgens, Hartmut, Heinz-Otto Peitgen, and Dietmar Saupe. *Chaos: Bausteine der Ordnung*. Berlin: Springer, 1994.

Kaiser, David. "Stick-Figure Realism: Conventions, Reifications, and the Persistence of Feynman Diagrams, 1948–1964." *Representations* 70 (2000): 49–86.

Kammerer, Paul. *Das Gesetz der Serie. Eine Lehre von den Wiederholungen im Lebens- und Weltgeschehen.* Stuttgart: Deutsche Verlags-Anstalt, 1919.

Kant, Immanuel. *Kritik der reinen Vernunft.* Riga: Hartknoch, 1787.

Kemp, Martin. *Visualizations: The Nature Book of Art and Science.* Oxford: Oxford University Press, 2000.

Kemp, Wolfgang. "Disegno. Beiträge zur Geschichte des Begriffs zwischen 1547 und 1607." *Marburger Jahrbuch für Kunstwissenschaft* 19 (1974): 219–40.

———. "Mittelalterliche Bildsysteme." *Marburger Jahrbuch für Kunstwissenschaft* 22 (1989): 121–34.

Kennedy, Barbara A. *Inventing the Earth: Ideas on Landscape Development since 1740.* Malden, MA: Blackwell, 2006.

Kepes, György. *The New Landscape in Art and Science.* Chicago: Paul Theobald, 1956.

———, ed. *The Structure in Art and in Science.* New York: Braziller, 1965.

———, ed. *Module, Proportion, Symmetry, Rhythm.* New York: Braziller, 1966.

Kleinman, Kent, and Leslie Van Duzer, eds. *Rudolf Arnheim: Revealing Vision.* Ann Arbor: University of Michigan Press, 1996.

Knopp, Konrad. *Mathematik und Kultur.* Berlin: W. de Gruyter, 1985.

Knorr-Cetina, Karin D. *Wissenskulturen. Ein Vergleich naturwissenschaftlicher Wissensformen.* Frankfurt am Main: Suhrkamp, 2002.

Koss, Juliet. *Modernism after Wagner.* Minneapolis: University of Minnesota Press, 2010.

Kra, Irwin, and Bernard Maski, eds. *Riemann Surfaces and Related Topics: Proceedings of the 1978 Stony Brook Conference.* Princeton: Princeton University Press, 1981.

Krantz, Steven G. "Fractal Geometry." *The Mathematical Intelligencer* 11 (1989): 12–16.

Krimmel, Bernd, ed. *Symmetrie in Kunst, Natur und Wissenschaft.* Exh. cat. Darmstadt, Germany: Mathildenhöhe, 1986.

Kuhn, Thomas S. *The Structure of Scientific Revolutions.* Chicago: University of Chicago Press, 1962.

Kunzle, David. *The History of the Comic Strip: The Nineteenth Century.* Berkeley: University of California Press, 1990.

Latour, Bruno. "Drawing Things Together." In *Representation in Scientific Practice.* Edited by Michael Lynch and Steve Woolgar, 19–68. Cambridge, MA: MIT Press, 1990.

———. *Pandora's Hope: Essays on the Reality of Science Studies.* Cambridge, MA: MIT Press, 1999.

Lawler, Gregory F., Oded Schramm, and Wendelin Werner. "The Dimension of the Planar Brownian Frontier is 4/3." In Citebase, http://arxiv.org/abs/math/0010165v2.

Lee, Jennifer B., and Miriam Mandelbaum. *Seeing Is Believing: 700 Years of Scientific and Medical Illustration.* New York: New York Public Library, 1999.

Letellier, Christophe, and Valérie Messager. "Influences on Otto E. Rössler's Earliest Paper On Chaos." *International Journal of Bifurcation and Chaos* 20, no. 11 (2010): 1–32.

Lévy, Paul. *Processus stochastiques et mouvement brownien.* Paris: Gauthier-Villars, 1948.

Lorenz, Edward N. "Deterministic Nonperiodic Flow." *Journal of the Atmospheric Sciences* 20 (1963): 130–41.

———. *The Essence of Chaos.* Seattle: University of Washington Press, 1993.

Lorenz, Konrad. *Der Abbau des Menschlichen. Die Rückseite des Spiegels. Versuch einer Naturgeschichte des Erkennens.* Munich: R. Piper, 1988.

Lovejoy, Shaun, and Benoît Mandelbrot. "Fractal Properties of Rain, and a Fractal Model." *Tellus Series A* 37 (1985): 209–32.

Lynn, Greg. *Folds, Bodies, and Blobs: Selected Essays*. Brussels: La Lettre Volée, 1998.

———. *Animate Form*. New York: Princeton Architectural Press, 1999.

———. *Folding in Architecture*. Chichester, UK: John Wiley & Sons, Ltd., 2004.

Lyotard, Jean-François. *La Condition postmoderne: rapport sur le savoir*. Paris: Les Éditions de Minuit, 1979.

Macho, Thomas. "Das Rätsel der vierten Dimension." In *Science & Fiction. Über Gedankenexperimente in Wissenschaft, Philosophie und Literatur*. Edited by Thomas Macho and Annette Wunschel, 62–80. Frankfurt am Main: Fischer, 2004.

Mandelbrot, Benoît. "Contribution à la théorie mathématique des jeux de communication." PhD diss., l'Université de Paris, 1953.

———. "On Recurrent Noise Limiting Coding." In vol. 3 of *Proceedings of the Symposium on Information Networks*, 205–22. Brooklyn, NY: Polytechnic Press, 1955.

———. "On the Theory of Word Frequencies and on Related Markovian Models of Discourse." In *Structures of Language and Its Mathematical Aspects*. Edited by Roman Jakobson, 190–219. Providence, RI: American Mathematical Society, 1961.

———. "Sur certains prix spéculatifs: faits empiriques et modèle basé sur les processus stables additifs de Paul Lévy." *Comptes Rendus* (Paris) 254 (1962): 2968–3970.

———. "New Methods in Statistical Economics." *The Journal of Political Economy* 71 (1963): 421–40.

——— "Self-Similar Error Clusters in Communication Systems and the Concept of Conditional Stationarity." *IEEE Transaction on Communications Technology* 13 (1965): 71–90.

———. "How Long Is the Coast of Britain? Statistical Self-Similarity and Fractional Dimension." *Science* 156 (1967): 636–38.

———. "Le syndrome de la variance infinie et ses rapports avec la discontinuité des prix." *Economie Appliquée* 26 (1973): 349–65.

———. *Les objets fractals: forme, hasard et dimension*. Paris: Flammarion, 1975.

———. "On the Geometry of Homogeneous Turbulence, with Stress on the Fractal Dimension of the Iso-Surfaces of Scalars." *Journal of Fluid Mechanics* 72 (1975): 401–16.

———. "Intermittent Turbulence and Fractal Dimension: Kurtosis and the Spectral Exponent 5/3+B." In *Turbulence and Navier Stokes Equations*. Edited by Roger Temam, 121–45. New York: Springer, 1976.

———. *Fractals: Form, Chance, and Dimension*. San Francisco: Freeman, 1977.

———. "Fractal Aspects of the Iteration of $z \to \lambda z\,(1\text{-}z)$ for complex λ and z." In *Nonlinear Dynamics: Annals of the New York Academy of Sciences* 357. Edited by Robert H. G. Helleman, 249–325. New York: New York Academy of Sciences, 1980.

———. "Fractals and the Geometry of Nature." In *1981 Yearbook of Science and the Future*. Chicago: Encyclopedia Britannica, Inc., 1980.

———. "Scalebound or Scaling Shapes: A Useful Distinction in the Visual Arts and in the Natural Sciences." *Leonardo* 14 (1981): 45–47.

———. "Des monstres de Cantor et de Peano à la géométrie fractale de la nature." In *Penser les mathématiques*. Edited by Jean A. Dieudonné, Françios Guénard, and Gilbert Lelièvre, 226–51. Paris: Seuil, 1982.

———. *The Fractal Geometry of Nature*. New York: Freeman, 1982.

———. "Les fractals, les monstres et la beauté." *Le débat: histoire, politique, société* 24 (1983): 54–72.

———. "Comment j'ai découvert les fractales." *La Recherche* 17 (1986): 420–24.

———. "Fractals and the Rebirth of Iteration Theory." In *The Beauty of Fractals: Images of Complex Dynamical Systems*. Edited by Heinz-Otto Peitgen and Peter H. Richter. Berlin: Springer, 1986

———. "Towards a Second Stage of Indeterminism in Science." *Interdisciplinary Science Reviews* 12 (1987): 117.

———. *Die fraktale Geometrie der Natur*. Basel and Boston: Birkhäuser, 1987.

———. "People and Events Behind the 'Science of Fractal Images.'" In *The Science of Fractal Images*. Edited by Heinz-Otto Peitgen, Dietmar Saupe, and M. F. Barnsley, 1–17. New York: Springer, 1988.

———. "Fractals and an Art for the Sake of Science." *Computer Art in Context: SIGGRAPH '89 Art Show Catalogue*, supplementary issue of *Leonardo* (1989): 21–24.

———. "Fractals and the Rebirth of Experimental Mathematics." In *Fractals for the Classroom: Part One: Introduction to Fractals and Chaos*. Edited by Hartmut Jürgens, Heinz-Otto Peitgen, and Dietmar Saupe, 1–16. New York: Springer, 1992.

———. "Les inattendus des fractales." *Pour la Science* 234 (1997): 10–12.

———. *Gaussian Self-Affinity and Fractals*. New York: Springer, 2002.

———. *Fractals and Chaos: The Mandelbrot Set and Beyond*. New York: Springer, 2004.

Marder, Michael. "Roughing It." *Science* 277 (1997): 647.

Markov, A. A. "Primer statističeskogo issledovanija nad tekstom 'Evgenija Onegina,' illyustrirujuščij svjaz' ispytanij v zep." *Izv. Imp. Akad. Nauk, Sankt Peterburg*, ser. 8, vol. 7, no. 3, (1913): 153–62.

———. "Ob odnom primenenie statističeskogo metoda," *Izv. Imp. Akad. Nauk, Sankt Peterburg,* ser. 6, vol. 10, no. 4 (1916): 239–42.

———. *The Correspondence Between A. A. Markov and A. A. Chuprov on the Theory of Probability and Mathematical Statistics.* Edited by Kh. O. Ondar. New York: Springer, 1981.

Mashaal, Maurice. *Bourbaki. Une société secrète de mathématiciens.* Paris: Belin, Pour la science, 2002.

Mead, Christopher. *Charles Garnier's Paris Opera: Architectural Empathy and the Renaissance of French Classicism.* Cambridge, MA: MIT Press, 1991.

Mehring, Christine. "Alma Buscher: 'Ship' Building Toy, 1923." In *Bauhaus 1919–1933: Workshops for Modernity.* Edited by Barry Bergdoll and Leah Dickerman, exh. cat., 156–59. New York: Museum of Modern Art, 2009.

Mehrtens, Herbert. *Moderne—Sprache—Mathematik: Eine Geschichte des Streits um die Grundlagen der Disziplin und des Subjekts formaler Systeme.* Frankfurt am Main: Suhrkamp, 1990.

———. "Mathematical Models." In *Models: The Third Dimension of Science.* Edited by Soraya de Chadarevian and Nick Hopwood, 276–306. Palo Alto, CA: Stanford University Press, 2004.

Meynen, Gloria. "Die Insel als Kulturtechnik (Ein Entwurf)." *Zeitschrift für Medienwissenschaft* 2 (2010): 79–91.

Miller, Wallis. "Architecture, Building, and the Bauhaus." In *Bauhaus Culture: From Weimar to the Cold War.* Edited by Kathleen James-Chakraborty, 74–75. Minneapolis: University of Minnesota Press, 2006.

Mitchell, William J. T. "The Pictorial Turn." *Artforum* 30 (1992): 89–94.

Moholy-Nagy, Sibyl. Review of *The New Landscape in Art and Science,* by György Kepes. *College Art Journal* 18 (1959): 192–94.

Molderings, Herbert. *Duchamp and the Aesthetics of Chance: Art as Experiment.* New York: Columbia University Press, 2010.

Morozov, N. A. *Die Evolution der Materie auf den Himmelskörpern: Eine theoretische Ableitung des periodischen Systems.* Dresden: Steinkopff, 1910.

———. *Die Offenbarung Johannis: eine astronomsich-historische Untersuchung.* Stuttgart: Spemann, 1912.

———. "Lingvističeskie spektry. Sredstvo dlja otličenija plagiatov ot istinnych proizvedenij togo ili drugogo izvestnogo avtora." *Isv. Akad. Nauk, otd. Russ. Jaz., Sankt Peterburg,* vol. 20 (1915): 93–134.

———. *Povest'moej žizni.* Moscow: Izd-vo Akademii nauk SSSR, 1961.

Most, Glenn W. *Doubting Thomas.* Cambridge, MA: Harvard University Press, 2005.

Muxel, Andreas, Martin Nawrath, Lasse Scherffig, and Georg Trogemann. "Experimente." In *Code und Material—Exkursionen ins Undingliche.* Edited by Georg Trogemann, 29–47. Vienna: Springer, 2010.

Nake, Frieder. *Ästhetik als Informationsverarbeitung. Grundlagen und Anwendungen der Informatik im Bereich ästhetischer Produktion und Kritik.* Vienna: Springer, 1974.

———. "Das doppelte Bild." *Bildwelten des Wissens, Kunsthistorisches Jahrbuch für Bildkritik* 3 (2005): 40–50.

Nierhoff, Barbara. *Georg Nees, Künstliche Kunst: Die Anfänge.* Bremen: Kunsthalle, 2005.

———. "Ex Machina—Die Begegnung von Computer und Kunst. Ein Blick zurück." In *Ex Machina—Frühe Computergrafik bis 1979.* Edited by Wulf Herzogenrath, Barbara Nierhoff, and Herbert W. Franke, 20–57. Berlin: Deutscher Kunstverlag, 2007.

Nietzsche, Friedrich. *Jugendschriften 1861–1864.* Vol. 2 of *Werke und Briefe. Historisch-kritische Gesamtausgabe.* Edited by Hans Joachim Mette. Munich: Deutscher Taschenbuch Verlag, 1994.

———. *Schreibmaschinentexte: Vollständige Edition, Faksimiles und kritischer Kommentar.* Edited by Stephan Günzel and Rüdiger Schmidt-Grépály. Weimar: Bauhaus-Universität Weimar, Universitätsverlag, 2002.

Nordmann, Alfred. "Vor-Schrift—Signaturen der Visualisierungskunst." In *Ästhetik der Wissenschaft. Interdisziplinärer Diskurs über das Gestalten und Darstellen von Wissen.* Edited by Wolfgang Krohn, 117–29. Hamburg: F. Meiner, 2006.

Ovenden, Mark. *Transit Maps of the World: The World's First Collection of Every Urban Train Map on Earth.* London: Penguin, 2007.

Panofsky, Erwin. *Perspective as Symbolic Form.* New York: Zone Books, 2002.

Pearson, Karl. "The Problem of the Random Walk." *Nature* 72 (1905): 294.

Peitgen, Heinz-Otto, and Peter Richter. *The Beauty of Fractals: Images of Complex Dynamic Systems.* Berlin: Springer, 1986.

Peitgen, Heinz-Otto, Hartmut Jürgens, and Dietmar Saupe. *Chaos and Fractals: New Frontiers of Science.* New York: Springer, 1989.

Pennisi, Elizabeth. "The Case of the Midwife Toad: Fraud or Epigenetics?" *Science* 325 (2009): 1194–95.

Perrig, Alexander. "Masaccios 'Trinità' und der Sinn der Zentralperspektive." *Marburger Jahrbuch für Kunstwissenschaft* 21 (1986): 11–43.

Perrin, Jean. "Mouvement Brownien et Molécules." *Journal de Physique Théorique et Appliquée* 9 (1910): 5–39.

Peterson, Ivars. "Fractal Past, Fractal Future." *Science News* 151 (1997): 13.

Pitt Rivers, Augustus H. Lane-Fox. "On the Principles of Classification Adopted in the Arrangement of His Anthropological Collection, Now Exhibited in the Bethnal Green Museum." *Journal of the Anthropological Institute of Great Britain and Ireland* 4 (1875): 293–308.

Plato. *Republic*. In *The Collected Dialogues of Plato*. Edited by Edith Hamilton and Huntington Cairns, translated by Paul Shorey, 575–845. New York: Pantheon, 1989.

Poincaré, Henri. *Science et Méthode*. Paris: Ernst Flammarion, 1908.

Prange, Regine. *Das Kristalline als Kunstsymbol: Bruno Taut und Paul Klee*. New York: Georg Olms, 1991.

Pratschke, Margarete. "Bildanordnungen." In *Das Technische Bild. Kompendium zu einer Stilgeschichte wissenschaftlicher Bilder*. Edited by Horst Bredekamp, Birgit Schneider, and Vera Dünkel, 116–19. Berlin: Akademie, 2008.

——. "Charles and Ray Eames' 'Powers of Ten'—Die künstlerische Bildfindung des Atoms zwischen spielerischem Entwurf und wissenschaftlicher Affirmation." In *Atombilder. Ikonografien des Atoms in Wissenschaft und Öffentlichkeit des 20. Jahrhunderts*. Edited by Charlotte Bigg and Jochen Hennig, 21–30. Göttingen: Wallstein, 2009.

——. "Windows als Tableau. Zur Bildgeschichte grafischer Benutzeroberflächen." PhD. diss. Humbolt-Universität, Berlin, 2010.

Prün, C. "G. K. Zipf's Conception of Language as an Early Prototype of Synergetic Linguistics." *Journal of Quantitative Linguistics* 6 (1999): 78–84.

Read, Herbert. *Preface to Aspects of Form: A Symposium on Form in Nature and Art*. Edited by Lancelot Law Whyte. London: Lund Humphries, 1951.

Reichler, Claude. *La découverte des Alpes et la question du paysage*. Geneva: Georg Editeur, 2002.

Rheinberger, Hans-Jörg. "Scrips and Scribbles." *MLN* 118 (2003): 622–36.

——. *Experimentalsysteme und epistemische Dinge: eine Geschichte der Proteinsynthese im Reagenzglas*. Frankfurt am Main: Suhrkamp, 2006.

Rheinberger, Hans-Jörg, Michael Hagner, and B. Wahrig-Schmidt, eds. *Räume des Wissens. Repräsentation, Codierung*, Spur. Berlin: Akademie, 1997.

Richardson, Lewis Fry. "The Problem of Contiguity: An Appendix to Statistics of Deadly Quarrels." In *General Systems: Yearbook of the Society for the Advancement of General Systems Theory* 6 (1961): 139–87.

Ritter, Carl. "Ueber räumliche Anordnungen auf der Aussenseite des Erdballs und ihre Functionen im Entwicklungsgange der Geschichten." In *Einleitung zur allgemeinen vergleichenden Geographie, und Abhandlungen zur Begründung einer mehr wissenschaftlichen Behandlung der Erdkunde*. Edited by Carl Ritter, 206–46. Berlin: Georg Reimer, 1852.

Rocker, Ingeborg M. "Calculus-Based Form: An Interview with Greg Lynn." *Architectural Design* 76 (2006): 88–95.

Rorschach, Hermann. *Psychodiagnostik. Methodik und Ergebnisse eines wahrnehmunsdiagnostischen Experiments (Deutenlassen von Zufallsformen)*. Edited by Walter Morgenthaler. Bern: Hans Huber, 1972.

Rosen, Margit. "Die Maschinen sind angekommen. Die [Neuen] Tendenzen—visuelle Forschung und Computer." *bit international. [Nove] tendencije—Computer und visuelle Forschung. Zagreb 1961–1973*. Edited by Margit Rosen and Peter Weibel, 31–57. Graz, Austria: Neue Galerie, 2007.

Rosen, Margit, and Peter Weibel. "Die Zukunft eines künstlerischen Mediums." In *Ex Machina—Frühe Computergrafik bis 1979*. Edited by Wulf Herzogenrath, Barbara Nierhoff, and Herbert W. Franke, 182–210. Berlin: Deutscher Kunstverlag, 2007.

Rössler, Otto E. "Chaotic Behavior in Simple Reaction Systems." *Zeitschrift für Naturforschung 31a* (1976): 259–64.

——. "Chaos in Abstract Kinetics: Two Prototypes." *Bulletin of Mathematical Biology* 39 (1977): 275–89.

——. "Chaos, Hyperchaos and the Double-Perspective." In *The Chaos Avant-Garde: Memories of the Early Days of Chaos Theory*. Edited by Ralph Abraham and Yoshisuke Ueda, 209–19. Singapore: World Scientific, 2000.

Rudwick, Martin. "The Emergence of Visual Language for Geological Science 1760–1840." *History of Science* 14 (1976): 149–95.

——. *Scenes from Deep Time: Early Pictorial Representations of the Prehistoric World*. Chicago: University of Chicago Press, 1992.

——. *Bursting the Limits of Time: The Reconstruction of Geohistory in the Age of Revolution*. Chicago: University of Chicago Press, 2005.

Ruskin, John. *The Elements of Drawing: In Three Letters to Beginners*. New York: Smith, Elder, & Co., 1857.

Samuel, Nina. "Form und Farbe digitaler Mathematik: Vom Zusammenspiel von zeichnender Hand und Computer in fraktalen Bildwelten." In *Bildwelten des Wissens, Kunsthistorisches Jahrbuch für Bildkritik* 3 (2005): 18–31.

——. "'I Look, Look, Look, and Play with Many Pictures.' Zur Bilderfrage in Benoît Mandelbrots Werk." In *Verwandte Bilder. Die Fragen der Bildwissenschaft*. Edited by Ingeborg Reichle, Steffen Siegel, and Achim Spelten, 297–320. Berlin: Kadmos, 2007.

——. "'Do Not Clean Off the Dust Specks. They Are Real.' Über gestörte, verschmutzte und verborgene Computerbilder." In *Prekäre Bilder*. Edited by Robert Suter and

Thorsten Bothe, 19–47. Munich: Wilhelm Fink, 2009.

———. "Beredte Skizzen. Chaos und das zerbrochene Ei von Yoshisuke Ueda." *Bildwelten des Wissens, Kunsthistorisches Jahrbuch für Bildkritik* 7, no. 2 (2010): 83–89.

———. "Die Form des Chaos. Bild und Erkenntnis in der komplexen Dynamik und der fraktalen Geometrie." PhD. diss., Humboldt-Universität, Berlin, 2011.

de Saussure, Horace-Bénédict. *Voyages dans les Alpes, précédés d'un essai sur l'histoire naturelle des environs de Genève.* Neuchâtel: Samuel Fauche, 1796.

Schlich, Thomas. "Repräsentationen von Krankheitserregern. Wie Robert Koch Bakterien als Krankheitsursache dargestellt hat." In *Räume des Wissens. Repräsentation, Codierung, Spur.* Edited by Hans-Jörg Rheinberger, Michael Hagner, and B. Wahrig-Schmidt, 165–90. Berlin: Akademie, 1997.

Schmidt-Burckhardt, Astrit. "Shaping Modernism: Alfred Barr's Genealogy of Art." *Word & Image* 16 (2000): 387–400.

Sekula, Allan. "The Body and the Archive." In *The Body: A Reader.* Edited by Mariam Fraser, 163–66. London: Routledge, 2005.

Senechal, Marjorie. "The Continuing Silence of Bourbaki: An Interview with Pierre Cartier." *The Mathematical Intelligencer* 20 (1998): 22–28.

Serres, Michel. *Über Malerei. Vermeer—La Tour–Turner.* Dresden: Verlag der Kunst, 1995.

Shannon, Claude E., and Warren Weaver. *The Mathematical Theory of Communication.* Urbana: University of Illinois Press, 2009.

Siegert, Bernhard. *Passage des Digitalen. Zeichenpraktiken der neuzeitlichen Wissenschaften 1500–1900.* Berlin: Brinkmann & Bose, 2003.

Sloan, Kim. "A New Chronology for Alexander Cozens Part II: 1759–86." *Burlington Magazine* 127 (1985): 355–63.

Smale, Steve. "Differentiable Dynamical Systems." *Bulletin of the American Mathematical Society* 73 (1967): 747–817.

Sobchack, Vivian. "A Theory of Everything: Meditations On Total Chaos." *Artforum* 19 (1990): 148–55.

Sørensen, Dan Erik Krarup. "Infinitely Renormalizable Quadratic Polynomials, with Non-Locally Connected Julia Sets." *The Journal of Geometric Analysis* 10, (2000): 169–206.

Speaks, Michael. "Chaos, Simulation, and Corporate Culture." *Mississippi Review* 17 (1989): 159–76.

Stafford, Barbara Maria. *Visual Analogy: Consciousness as the Art of Connecting.* Cambridge, MA: MIT Press, 1999.

Touret, Jaques. "Élie de Beaumont (1798–1874), des systèmes de montagnes au réseau pentagonal." In *L'essor de la géologie française.* Edited by Jean Gaudant, 95–132. Paris: Presses des Mines, 2009.

Tufte, Edward R. *The Visual Display of Quantitative Information.* Cheshire, CT: Graphics Press, 2002.

———. *Beautiful Evidence.* Cheshire, CT: Graphics Press, 2007.

Vargas, Alexander O. "Did Paul Kammerer Discover Epigenetic Inheritance? A Modern Look at the Controversial Midwife Toad Experiments." *Journal of Experimental Zoology* 312 B (2009): 667–78.

Vasari, Giorgio. *Vasari on Technique.* Translated by Louisa S. Maclehose. London: Dent, 1907.

Velminski, Wladimir. "Der Speck am Text. Die Entstehung der Genetik aus der Berechenbarkeit der Literatur." In *Andrej A. Markov. Berechenbare Künste. Mathematik, Poesie, Moderne.* Edited by Ph. von Hilgers and Wladimir Velminski, 101–26. Zurich: diaphanes, 2007.

Venturi, Robert. *Complexity and Contradiction in Architecture.* New York: The Museum of Modern Art, 1966.

Viollet-le-Duc, Eugène Emmanuel. *Le Massif du Mont-Blanc. Étude sur sa construction géodésique et géologique, sur ses transformations, et sur l'état ancien et moderne de ses glaciers.* Paris: Baudry, 1876.

Volkert, Klaus Thomas. *Die Krise der Anschauung: Eine Studie zu formalen und heuristischen Verfahren in der Mathematik seit 1850.* Göttingen, Germany: Vandenhoeck & Ruprecht, 1986.

von Brevern, Jan. *Blicke von Nirgendwo: Geologie in Bildern bei Ruskin, Viollet-le-Duc, und Civiale.* Munich: Wilhelm Fink, 2011.

Wainer, Howard. "Graphical Visions from William Playfair to John Tukey." *Statistical Science* 5 (1990): 340–46.

Walsh, J. L. "Another Contribution to the Rapidly Growing Literature of Mathematics and Human Behavior." *Scientific American* 181 (1949): 56–58.

Warnke, Martin. *Political Landscape: The Art History of Nature.* London: Reaktion Books, 1994.

———. ed. *Politische Kunst.* Berlin: Akademie, 2004.

Wasmund, Klaus. *Politische Plakate aus dem Nachkriegsdeutschland. Zwischen Kapitulation und Staatsgründung 1945–1949.* Frankfurt am Main: Fischer, 1986.

Wellmann, Janina. *Die Form des Werdens: Eine Kulturgeschichte der Embryologie, 1760–1830.* Göttingen: Wallstein, 2010.

Wells, H. G. *The Island of Dr. Moreau.* Garden City, NY: Garden City Publishing Co., 1896.

Weltzien, Friedrich. *Fleck, das Bild der Selbsttätigkeit: Justinus Kerner und die Klecksografie als Experimen-*

telle Bildpraxis zwischen Ästhetik und Naturwissenschaft. Göttingen: Vandenhoeck & Ruprecht, 2011.

Werner, Gabriele. "Das Bild vom Wissenschaftler—Wissenschaft im Bild. Zur Repräsentation von Wissen und Autorität im Portrait am Ende des 19. Jahrhunderts." *kunsttexte.de* 1 (2001): 1–11.

Weyl, Hermann. *Gesammelte Abhandlungen*. Edited by Komaravolu Chandrasekharan. Berlin: Springer, 1968.

Whitehead, Alfred North. *Science and the Modern World*. Cambridge: Cambridge University Press, 1926.

Whyte, Lancelot Law, ed. *Aspects of Form: A Symposium on Form in Nature and Art*. London: Lund Humphries, 1951.

Wille, Rudolf ed., *Symmetrie in Geistes- und Naturwissenschaft: Hauptvorträge und Diskussionen des Symmetrie-Symposiums an der Technischen Hochschule Darmstadt vom 13. bis 17. Juni 1986 im Rahmen des Symmetrieprojektes der Stadt Darmstadt*. Berlin: Springer, 1988.

Wind, Edgar. *Art and Anarchy*. London: Faber and Faber, 1963.

Wright, Richard. "Art and Science in Chaos: Contested Readings of Scientific Visualizations." In *FutureNatural: Nature, Science, Culture*. Edited by George Robertson, 218–36. London: Routledge, 1996.

Wullen, Moritz, ed. *Natur als Vision: Meisterwerke der englischen Präraffaeliten*. Berlin: SMB-DuMont, 2004.

Ziegler, Günter M. "Countdown. Über Mathematik und Science Fiction." In *Gegenworte. Hefte für den Disput über Wissen Bilder (in) der Wissenschaft* 20 (2008): 73–82.

Zipf, George Kingsley. *Selected Studies of the Principle of Relative Frequency in Language*. Cambridge, MA: Harvard University Press, 1932.

———. *The Psycho-Biology of Language: An Introduction to Dynamic Philology*. Cambridge, MA: Harvard University Press, 1935.

———. *Human Behavior and the Principle of Least Effort: An Introduction to Human Ecology*. Cambridge, MA: Harvard University Press, 1949.

Index

Photographic Credits

Photographs were taken or supplied by the lending institutions, organizations, or individuals credited in the picture captions and are protected by copyright; many names are not repeated here. Individual photographers are credited below. Permission has been sought for use of all copyrighted illustrations in this volume. In several instances, despite extensive research, it has not been possible to locate the original copyright holder. Copyright owners of these works should contact the Bard Graduate Center, 18 West 86th Street, New York, NY, 10024.

Angela Broehan: page 151, fig. 5

Eric Edler: page 11, fig. 1 and page 12, fig. 2

Juliet Koss: page 148, fig. 1

Digital Image © The Museum of Modern Art/Licensed by SCALA / Art Resource, NY. © 2012 Artists Rights Society (ARS), New York / VG Bild-Kunst, Bonn: page 151, fig. 4

Bruce M. White: pages 112–13, cats. 17 and 18